PHYSICS AND MATHEMATICS
OF STRINGS

PHYSICS AND MATHEMATICS OF STRINGS

PROCEEDINGS OF
A ROYAL SOCIETY DISCUSSION MEETING
HELD ON 8 AND 9 DECEMBER 1988

ORGANIZED AND EDITED BY
SIR MICHAEL ATIYAH, F.R.S., J. R. ELLIS, F.R.S.,
M. B. GREEN AND C. H. LLEWELLYN-SMITH, F.R.S.

LONDON
THE ROYAL SOCIETY
1989

Printed in Great Britain for the Royal Society
by the
University Press, Cambridge

ISBN 0 85403 393 9

First published in *Philosophical Transactions of the Royal Society of London,*
series A, volume **329** (no. 1605), pages 317–413

Copyright

British Library Cataloguing in Publication Data

Physics and mathematics of strings.
1. High energy physics. String theory
I. Atiyah, *Sir*, Michael, *1929–*   II. Royal Society
539.7′2

ISBN 0-85403-393-9

Published by the Royal Society
6 Carlton House Terrace, London SW1Y 5AG

PREFACE

One of the most exciting developments in fundamental physics during the past few years has been the discovery that string theory apparently provides a viable framework for unifying elementary particles and their interactions. In particular, string theory enables one to calculate quantum gravitational effects that were infinite and meaningless in previous quantum field theories. The study of string theory involves deep issues in mathematics, and has close connections with the theory of phase transitions as well as other aspects of contemporary physics.

The Discussion Meeting held on 8 and 9 December 1988 at the Royal Society brought together many of the leading researchers in string theory and related areas of mathematics. The meeting covered many of the most interesting developments in the physical applications and mathematical structure of strings, as well as raising thought-provoking questions for the future. It provided a uniquely valuable forum for contacts between physicists and mathematicians working on related problems.

The topics covered range from profound meditations on the nature of space and time to attempts at describing the standard model of elementary particles in the framework of string theory. They include expositions by mathematicians of tools needed by physicists, as well as attempts by the latter to apply them.

The organizers are grateful to all speakers at the meeting for their stimulating presentations. They are also grateful for the timely production of the contributions collected in this volume. We gratefully acknowledge the enormous amounts of help provided by Miss C. A. Johnson in organizing the meeting and by Mr J. E. Wainwright in editing these proceedings.

August 1989

M. ATIYAH
J. R. ELLIS
M. B. GREEN
C. H. LLEWELLYN-SMITH

CONTENTS

Phil. Trans. R. Soc. Lond. A **329**, 319–328 (1989)

Printed in Great Britain

319

Introduction to string theory: its structure and its uses

By D. I. Olive, F.R.S.

Department of Theoretical Physics, Blackett Laboratory, Imperial College of Science and Technology, Prince Consort Road, London SW7 2BZ, U.K.

Over five years ago experiments at CERN confirmed that the weak (radioactive) interactions of elementary particles are mediated by gauge particles that are heavy relatives of the photon, namely the quantum of light and radio wave propagation. Gauge particles have to belong to a pattern given by the structure of a compact Lie group. Mathematicians listed such patterns at the beginning of the century and it seems that nature favours one of the 'exceptional' possibilities when nuclear forces are included.

Twenty years ago a picture of elementary particles as quantums of the excitations of a one-dimensional string was developed. Consistency with the principles of relativity and quantum mechanics seemed to require the aforementioned exceptional gauge structure as well as gravitational forces in Einstein's formulation. Thus a simple 'string' principle promised to explain and unify all the diverse fundamental forces of nature: electromagnetic, weak, nuclear and gravitational. Unfortunately, there remain detailed questions still to be resolved.

Nevertheless, the theory possesses rich mathematical structure encompassing Lie algebras and infinite-dimensional generalizations and complex algebraic geometry in a way which sheds valuable new perspectives on modern pure mathematics. At the same time it has unexpected applications in describing and classifying the modes of phase transition in two-dimensional materials, a classical problem in statistical physics.

An ideal introduction to string theory could, with some justice, be subtitled 'The string theory prerequisites for mathematics', but that would be too ambitious a title for what I have to say.

The reason I mention this is to draw attention to the fact that string theory is a theory of elementary particles and their fundamental interactions largely created in the decade after 1968 by physicists who were largely ignorant of modern advanced mathematics (some of which had not even been developed then). They certainly believed that they had a theory of considerable mathematical significance, but were unable to interest mathematicians in it at that time. Equally, they were unable to persuade the larger physics community of the relevance of their theories. Consequently, the theory went into hibernation only to reawaken dramatically when it was recognized that there was a rapprochement between string theory and newer ideas in particle physics, grand unification and the theory of anomalies, as well as ideas further afield, such as the representation theory of affine Kac–Moody algebras and the theory of Riemann surfaces and their moduli. Since then, progress has continued apace on several fronts, providing the occasion for this meeting.

String theory was, of course, always based on physical principles, but these have shifted as the structure has been clarified. As the original version was designed to include observed features of the scattering of nuclear particles, the physical principles were couched in terms of the scattering matrix and the presentation of the theory was algebraic in nature. Later it was

[1]

realized that the theory applied rather to the fundamental forces between leptons and quarks rather than the forces between neutrons and protons and as a consequence the basic principles and the presentation of the theory became more geometric in nature. The result of this was that the same basic theory could be formulated in two apparently different ways, algebraic and geometric, and it is the comparison between these two approaches which has related different branches of mathematics in unexpected and fascinating ways, as described by other contributors to this Symposium.

Historically, the graph of the development of string theory has resembled a staircase with seemingly insurmountable obstacles suddenly overcome by a new insight, usually based on the appreciation of an unexpected manifestation of symmetry in a form subtly different from hitherto, for example, the role of the Virasoro algebra, super versions of it, and affine Kac–Moody algebras.

In motivating string theory I shall proceed directly to the current point of view, ignoring the history that can be gleaned from the previous Royal Society Discussion Meeting in 1969 or my talk at the 1974 London conference on particle physics (Olive 1974).

Hitherto the twentieth century has seen four new physical principles:

  (i) Einstein's theory of special relativity (which unified space and time);

 (ii) quantum mechanics;

(iii) the gauge principle, i.e. invariance under internal symmetries performed independently at different points of space and time;

(iv) gravity according to Einstein's theory of general relativity.

The simplest example of (iii) is a change in phase of the Schrödinger wave function for an electron;

$$\psi(x, t) \rightarrow e^{iq\chi(x,\, t)/\hbar}\, \psi(x, t). \tag{1}$$

It is a familiar feature of quantum mechanics that wave functions are complex, with an overall phase that is irrelevant. The phase change that occurs in (1) varies from point to point in space and time. Such phase changes evidently form a $U(1)$ group at each point of space and time, called the gauge group. The demand that physics be independent of these 'gauge rotations' leads to the introduction of a 'gauge potential' or 'connection' satisfying Maxwell's equations, and hence the phenomena of electricity and magnetism, together with light- and radio-wave propagation. The constant q in the phase change is the electric charge of the electron. The most natural generalization of this group $U(1)$ of unitary one by one matrices is to the group $U(2)$ of unitary 2 by 2 matrices. The gauge theory of this yields a nonlinear generalization of Maxwell's equations satisfied by four gauge fields (corresponding to the four generators of the group $U(2)$). This successfully describes the phenomena of radioactivity in addition to electromagnetism, which corresponds to a $U(1)$ subgroup of $U(2)$ (not the invariant one). This is the Salam–Weinberg theory of electroweak interactions that was experimentally verified at CERN in 1982–83 by the detection of the three new gauge particles. The reason that detection was so difficult was that, owing to symmetry breaking effects, they were not massless like the photon, which is the particle of the $U(1)$ gauge potential, but heavier than a hydrogen atom.

The success of this theory suggested that the nuclear forces be included similarly. It had already been realized that the forces between the quarks, the constituents of the protons and neutrons, were of the gauge type with group $SU(3)$ (of 3 by 3 unitary matrices with unit determinant). The actual nuclear forces between the neutrons and protons were not of this type, being a complicated derived effect of the former. Because of the ubiquity of gauge forces an attractive goal would be a 'grand unification' whereby the gauge group was a simple Lie group containing $SU(3) \times U(2)$ as economically as possible. The most promising candidates turned out to be the following sequence of groups:

$$U(2) \times SU(3) \subset SU(5) \subset SO(10) \subset E_6 \subset E_7 \subset E_8. \tag{2}$$

The larger groups here are successively more approximate (and uncertain) owing to the symmetry-breaking effects. Nevertheless, once the matter particles, i.e. electrons, muons, neutrinos and quarks, are assigned to representations of these groups, usually what Bourbaki calls miniscule representations, we have an extremely succinct encapsulation of thousands of millions of pounds of experimental data. However, these statements are not the whole story: there are the symmetry-breaking effects to be studied in more detail by the future accelerators and the theoretical question of how to include gravity (iv). The results (2) add the stunning new question as to why nature has an apparent predilection for exceptional structures such as the E groups. The interest in string theory is that it promises successful answers to the latter two questions even though the question of symmetry breaking is still beyond it.

Not all theories of the gauge type are internally consistent when quantum mechanics is fully taken into account. The famous results from the late 1950s concerning the non-conservation of parity in weak (radioactive) interactions showed that left- and right-handed matter should be treated differently, as they transform under inequivalent representations of the grand unified gauge group (actually complex conjugates of each other in four dimensions of space and time, and hence complex). The trouble is that it is not always possible to quantize a given classical theory while preserving a symmetry because of what are called 'anomalies'. This can be seen from quantum mechanical action principle of Richard Feynman (1948):

$$\text{matrix element} = \iint \delta \, \text{fields} \, e^{i\,\text{action}/\hbar}. \tag{3}$$

In a conventional field theory the action is an integral over space-time of the lagrangian density that depends only on the fields and their derivatives evaluated at the relevant space-time point. In the classical limit of Planck's constant h being very small this gives the principle of stationary action, as it should, with local field equations. Symmetry in classical physics is guaranteed if the action is invariant, but the above shows that in quantum theory the integration measure over the fields must also be invariant. This is not automatic in the case of gauge symmetry when left-handed and right-handed matter transform according to inequivalent representations of the gauge group. (This effect is closely related to the Atiyah–Singer index theorem.) For the grand unified theories in four dimensions of space and time this problem is fortunately evaded by the choice of representations assigned to matter on the basis of the data.

The problem of anomalies will reappear in various guises when we have dealt with the next difficulty, which is the occurrence of divergences in the above expression, particularly when gravity is taken into account following Einstein's theory. Because we are dealing with local field equations we are necessarily imagining that the elementary particles occupy isolated points in space. This picture is rather singular especially when interaction is included. As time evolves, the particle point follows a 'world-line', depicting its trajectory in space and time with interactions occurring at the junctions of these world-lines; see figure 1.

[3]

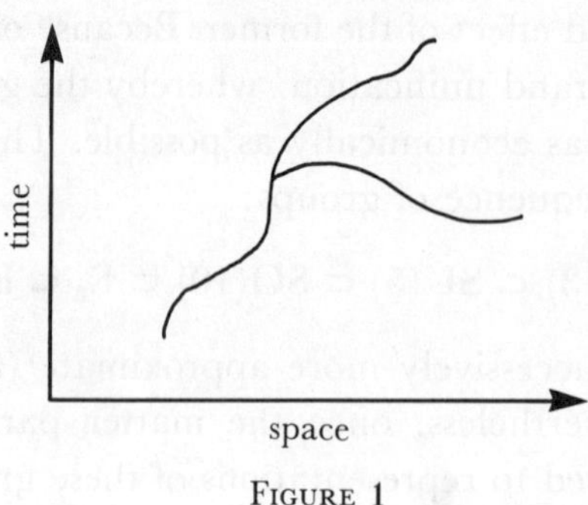

FIGURE 1

These junctions are highly singular points in this description and are the points where the divergences originate. Unfortunately, they appear to be an inevitable consequence of the principles (i)–(iii), which had, in other respects, been so successful. One lesson can be salvaged and that is that the action for freely moving point particles is beautifully simple and geometric:

$$\text{action} = -mc^2 \int d\tau, \tag{4}$$

where τ is the proper time of the particle, i.e. that measured by a clock travelling with it, and the integration is along the world-line of the particle. Thus the action is simply the length of the world-line and is stationary precisely for straight-line trajectories.

The string revolution started 20 years ago when it was realized that it was worth picturing elementary particles as occurring in families corresponding to the quantized modes of excitation of string. The original reason was to explain the spectrum of observed particles that included high spins, but later the motivation changed as we shall see. Thus if the string is open ended, I have as possible motions those depicted in figure 2.

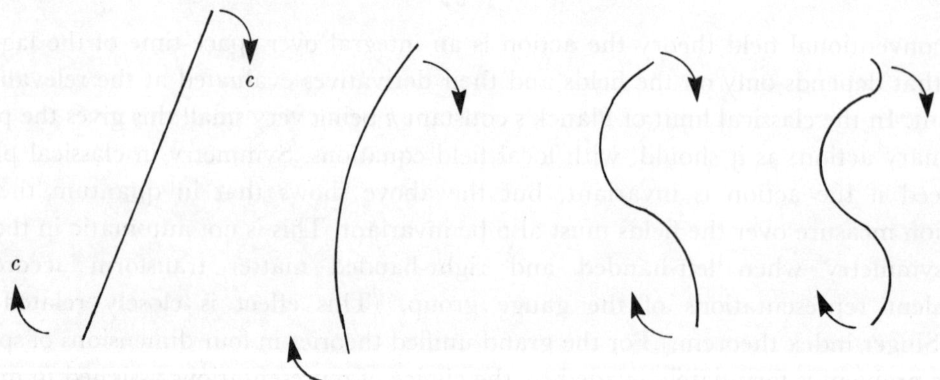

FIGURE 2

These illustrations indicate that the end-points of the string move at the speed of light. The frequency of the vibrations superimposed on the overall rotation are harmonics of a fundamental, just as they are for a vibrating violin string which has fixed end-points, even though the two boundary conditions are quite different.

These are classical solutions and so render stationary the action that again is extremely simple and geometric (Goto 1971; Y. Nambu, unpublished work). It is the two-dimensional 'area' of the world-sheet swathed out by the string as it moves in space and time. This is a natural generalization of the action (4), but, as we shall see, possesses more symmetry and intrinsic advantages compared with (4) in determining interactions:

$$\text{action} \propto \text{area of world-sheet} = \iint d\tau \, d\sigma \, \sqrt{(-\det h)}. \tag{5}$$

To say what we mean by area we need a metric on the world-sheet (supposing it to be a manifold) and we take the natural one imposed by its embedding in space and time that has the usual (flat) Lorentz metric $g^{\lambda\mu}$ appropriate to $D-1$ space variables and one time variable. Because of its geometric nature, the action (5) is independent of the choice of coordinates on the world-sheet. It is usual to suppose that the sheet possesses at least one tangent that is time-like at each point. $\tau = \tau^1$ is a time-like variable and $\sigma = \tau^2$ a space-like one. The coordinate in space-time of the point (τ, σ) is denoted $X^\lambda(\tau, \sigma)$. Then the metric is given by

$$h_{\alpha\beta} = \partial_\alpha X^\mu g_{\mu\nu} \partial_\beta X^\nu, \quad \text{where} \quad \partial_\alpha = \partial/\partial\tau^\alpha \quad \text{or} \tag{6a}$$

$$h = \begin{pmatrix} \dot{X}^2 & \dot{X}X' \\ \dot{X}X' & X'^2 \end{pmatrix} \tag{6b}$$

in matrix notation. Hence

$$-\det h = (\dot{X}X')^2 - \dot{X}^2 X'^2, \tag{7}$$

where dot and prime denote differentiations with respect to τ and σ, respectively.

There is an important feature of the action (5) that agrees with everyday experience of electromagnetic waves and so lends weight to the idea that (5) is the beginning of a theory encompassing the gauge theories previously discussed. This is called 'transversality'. The elementary particles are the quantized vibrations of the string harmonics previously mentioned. If we picture the vibrations in space-time, we see that the vibrations within the world-sheet have no meaning as they can be redefined away by a change of the variables σ and τ. This is good because we do not want vibrations in time. These would be difficult to interpret and lead to 'ghosts', which are incompatible with the principles of quantum mechanics. But it also follows that we can redefine away those space-like vibrations along the string, i.e. longitudinal vibrations. This leaves only $D-2$ meaningful directions transverse (or normal) to the world-sheet in which physical vibrations can occur.

When $D = 4$, our usual situation, this leaves two directions of transverse polarization or vibration. Light waves exhibit this feature by possessing two states of polarization orthogonal to the direction of propagation. This can be seen by superimposing two lenses of polarized sunglasses and rotating them relative to each other. Radio waves are also polarized transversally as is demonstrated by the fact that radio aerials on cars have to be mounted vertically.

This transversality plays an important role in the mathematical structure of string theory. It is related to the reparametrization invariance of the string action (5) and this recalls general covariance in Einstein's general relativity. I now explore the idea that the action (5) can be formulated in a way more like that theory.

[5]

From that point of view, I think of the X^μ as D scalar fields on the world-sheet furnished with an $\mathrm{SO}(D-1, 1)$ internal symmetry. This suggests I consider an alternative action to (5):

$$\text{action} = \iint d\sigma \, d\tau \, \sqrt{(-\det \tilde{h})} \, \partial_\alpha X^\mu g_{\mu\nu} \partial_\beta X^\nu \tilde{h}^{\alpha\beta} \tag{8a}$$

$$= \iint d\sigma \, d\tau \, \sqrt{(-\det \tilde{h})} \, h_{\alpha\beta} \tilde{h}^{\alpha\beta} \tag{8b}$$

by using ($6a$) for the embedded metric. As in general relativity, I now treat the dummy metric $\tilde{h}_{\alpha\beta}$ as an independent variable in the action, in addition to the X^λ, which were the previous independent variables. There are consequently two Euler–Lagrange equations resulting from rendering (8) stationary to variations first of $\tilde{h}$:

$$\theta_{\alpha\beta} = 0, \tag{9}$$

where

$$\theta_{\alpha\beta} = \partial_\alpha X^\mu g_{\mu\nu} \partial_\beta X^\nu - \tfrac{1}{2} \tilde{h}_{\alpha\beta} h^{\gamma\delta} \tilde{h}_{\gamma\delta}, \tag{10}$$

and then of X^λ

$$\Delta X^\lambda = \partial_\alpha (\sqrt{(-\det \tilde{h})} \, \tilde{h}^{\alpha\beta} \partial_\beta X^\lambda) = 0. \tag{11}$$

Expression (10) is interpreted as the energy–momentum tensor within the string world-sheet. By (9), it vanishes, expressing the fact that there is no observable flow of energy within the world-sheet. This is another statement of transversality. It also follows from (9) that the dummy and induced metrics $\tilde{h}$ and h are proportional. The action ($8b$) is independent of the overall scale in $\tilde{h}$ and, as a consequence, ($8b$) reduces to the Nambu–Goto action (5). The action (8) is usually known as the Polyakov action though it is not due to him. The second Euler–Lagrange equation (11) states that X^λ satisfies the covariant Laplace equation on the world-sheet. If τ and σ are chosen to be orthonormal in the sense that ($6b$) reads as

$$\begin{pmatrix} 1 & 0 \\ 0 & -1 \end{pmatrix},$$

this reduces to the wave equation, thus explaining the wave solutions described in figure 2.

The symmetry of the action (8) under rescaling of $\tilde{h}$, the dummy metric, is a new feature compared with the Nambu–Goto action (5). The process is called Weyl rescaling and henceforth assumed to be of fundamental importance. This new principle would forbid me adding a cosmological term $\int d\sigma \, d\tau \, \sqrt{(-\det \tilde{h})}$ to (8) but would allow an Einstein term

$$A(\text{Einstein}) = \text{const.} \iint d\sigma \, d\tau \, \sqrt{(-\det \tilde{h})} \, R, \tag{12}$$

where R is the scalar curvature. Because the world-sheet is two dimensional, the integrand in (12) is a total derivative. As a consequence the equations of motion already considered would be unaffected by the addition of (12) to (8).

However, it could have important global effects. So far, I have mentioned strings with ends but these imply that I must also have strings with no ends, i.e. closed strings. Their world-sheets have no boundary and are easier to consider. A freely propagating closed string would produce a world-sheet like a cylinder, see figure 3. But I could consider a surface with more complicated topology, as in figure 4, with g holes. Physically, this describes a string that splits

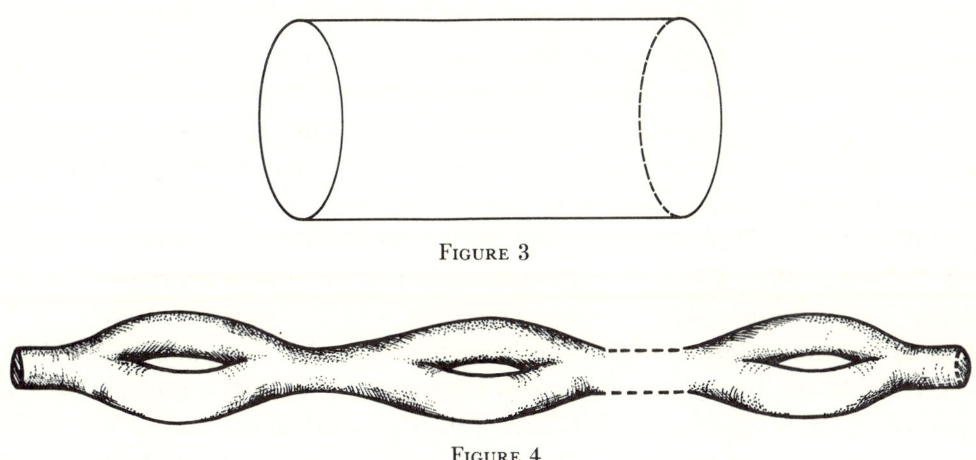

FIGURE 3

FIGURE 4

and recombines g times by means of $2g$ vertices. In this way the string manifests interaction. The contributions to the Feynman integral (3) from such surfaces all have a common factor

$$e^{iA(\text{Einstein})} = e^{i\,\text{const.}(\text{Euler number})} = e^{i\,\text{const.}(2-2g)},$$

by the Gauss–Bonnet theorem, as $A(\text{Einstein})$ is topological. Thus all world-sheets with the same topology contribute a common factor raised to the power of the number of interactions via the constant in (12). This is the interaction or coupling constant.

We see that string theory possesses a simple geometrical mechanism for describing interaction by means of a complication of the topology of the world-sheet through the addition of holes. This mechanism would seem unlikely to produce divergences because there is no apparent singular point of the type that we saw was responsible for the divergences in the point-particle situation. Yet the successes of the gauge theories could also be encompassed if we could understand the choice of gauge group.

The discussion so far is incomplete in that the absence of anomalies in quantum mechanics must be properly checked. This can be done in at least two different ways leading to the same conclusions: (1) the open string spectrum contains a massless spin one particle, i.e. a gauge particle, (2) the closed string spectrum contains a massless spin two particle, i.e. the graviton, the carrier of gravitational forces, and (3) the theory avoids anomalies in Lorentz covariance only if space and time has 26 dimensions!

The first two results are highly satisfactory in that they provide more evidence that string theory does indeed include the gauge and gravitational forces. Some greeted the third result with consternation while others regarded it as the first indication of exceptional structure. It was Lovelace (1971) who was the first to realize that there had to be $26-2 = 24$ transverse dimensions to facilitate the action of an element of the modular group that related closed string states to open string states. The exceptional structure of which this was suggestive is the special even self-dual lattice in 24 dimensions, the Leech lattice. Now we know that there is a holomorphic conformal field theory related to this whose symmetry group is the famous monster group (Frenkel *et al.* 1988).

The string theory described so far has the serious drawback that the particle states are all bosonic and never fermionic. Hence the theory lacks leptons and quarks, an essential ingredient for a would-be unified theory. A variant of string theory including fermions was initiated by

Ramond (1971) and by Neveu & Schwarz (1971) and developed in the years following. It possessed supersymmetry on the world-sheet (and stimulated much activity in that subject). For consistency it required 10 dimensions of space and time rather than 26, but, like the purely bosonic theory, possessed a particle moving faster than light and thus unacceptable. The fermionic string theory had the advantage that this tachyon could be eliminated in a way that not only respected the existing symmetries, but enlarged them to manifest supersymmetry in space and time (Gliozzi *et al.* 1977). As a result this theory was later called the 'superstring' (Green & Schwarz 1981). The construction of this supersymmetry algebra owed much to the fact that the transverse space has the same dimension, $10-2 = 8$, as the algebra of octonions. Indeed, the exceptional Jordan algebra of hermitian three by three matrices with octonion entries plays a role and I sense that the trail of exceptional structures is growing even warmer.

It is possible to add gauge symmetry to such a theory, but then one has to check the absence of an anomaly for it in 10 dimensions. There the representations carried by left-handed and right-handed fermionic matter are separately real and unrelated. In fact, by supersymmetry, one of these representations is trivial and the other adjoint to match the gauge particles. Green & Schwarz found that, given this, the anomaly cancelled only if the gauge group was $E_8 \times E_8$ or $SO(32)$. The easiest way to understand this dramatic result is to note that both these possess rank 16 and weight lattices (or sublattices) that are even and self-dual, properties good for constructing theta functions with simple behaviour under the action of the modular group (an important feature of a consistent string theory). The first possible choice of group is close to explaining the exceptional group structures observed in the sequence (2).

A superstring theory with this gauge symmetry was found by the 'heterotic' construction, so called because it awkwardly matched a 26-dimensional left-handed bosonic string with a 10-dimensional right-handed superstring (Gross *et al.* 1985). The 16 excess dimensions of the bosonic string moved on a maximal torus of either of the stated groups, thereby yielding that group as a gauge symmetry according to the vertex operator construction of Frenkel & Kac (1980) and of Segal (1981).

This leaves the problem of reducing from 10 to four dimensions, or of moving in some more direct manner with an explanation of the symmetry-breaking effects mentioned earlier. There is much activity, but as yet no satisfactory and compelling solution. Peter Goddard, Graham Ross and John Schwarz discuss the latest progress in their papers in this Symposium.

The question of divergences is still open, with no universally accepted proof of their absence. S. Mandelstam (unpublished work) has announced a proof for superstring theory but, as far as I know, no other workers have been able to examine the details of this proof yet. This result would be of considerable importance as it would establish, for the first time, a finite theory of gravity.

I can now say something about the quantization of the string motion. This can be done algebraically, forming the hamiltonian and using canonical quantization rules. An important role is played by the energy–momentum tensor (10), which is traceless in the sense that $h^{\alpha\beta}\theta_{\alpha\beta} = 0$, by definition, without recourse to (9). This tensor has only two components, which when appropriately chosen depend only on σ and τ, respectively. In a suitable basis these two components generate two commuting copies of the Virasoro algebra:

$$[L_m, L_n] = (m-n)\, L_{m+n} + \tfrac{1}{12} D\, m(m^2 - 1)\, \delta_{m+n,\,0}. \tag{13}$$

The non-zero term on the right-hand side of (13) means that (9) cannot hold quantum mechanically (unless Faddeev–Popov ghosts are introduced and $D = 26$). Instead one considers 'physical states' corresponding to physical particles satisfying

$$(L_m - \delta_{m,0})\,|\text{phys}\rangle = 0. \tag{14}$$

Because of time-like oscillations the Hilbert space of quantized oscillations is not positive definite. However, the subspace satisfying (14) is positive definite when a subspace of zero-norm states is modded out. This is the celebrated 'no ghost' theorem of Brower (1972) and of Goddard & Thorn (1972), valid providing D is less than or equal to 26. The structure of this proof has lead to a representation theory of the Virasoro algebra (13) for general values of D not necessarily an integer. This theory is used for classifying more general theories with conformal symmetry and in particular, gives the critical exponents that govern the power law behaviour of correlation functions in two-dimensional materials making second-order phase transitions (Belavin *et al.* 1984; Friedan *et al.* 1984). This exciting application of string theory has developed rapidly with affine Kac–Moody algebras playing a key role (Goddard & Olive 1986).

Finally, I briefly mention the more geometric approach to quantization associated with the name of Polyakov (1981). We have seen that the string action (8) is invariant under both diffeomorphisms and Weyl rescalings of the metric $\tilde{h}_{\alpha\beta}$. Thus in the integration over metrics implied by the Feynman integral (3), I should only integrate over conformal classes not so related. If, following Polyakov, I imagine performing a 'Wick rotation' in which time becomes imaginary so that both the space-time metric $g^{\lambda\mu}$ and the world-sheet metric $\tilde{h}^{\alpha\beta}$ become positive definite, the conformal classes now label Riemann surfaces. The integration in (3) is that over the moduli space of Riemann surfaces of a given topology (at least for closed strings) and is finite dimensional. However, to obtain this I have to divide out by the volume of the invariance group, a manoeuvre that is customarily performed in quantum field theory by the trick of introducing the Faddeev–Popov ghosts as extra fields (which relate to the cohomology of the Virasoro algebra (13)). The new, combined integration measure is anomaly-free only if $D = 26$ or 10, as appropriate. Obviously, the development of these ideas uses the theory of moduli spaces in an interesting way which is particularly intriguing when it is compared with the algebraic approach previously mentioned.

R E F E R E N C E S

Belavin, A. A., Polyakov, A. M. & Zamolodchikov, A. B. 1984 Infinite conformal symmetry in two-dimensional quantum field theory. *Nucl. Phys.* B **241**, 333–380.

Brower, R. 1972 Spectrum generating algebra and no-ghost theorem for the dual model. *Phys. Rev.* D **6**, 1655–1662.

Feynman, R. P. 1948 Space-time approach to relativistic quantum mechanics. *Rev. mod. Phys.* **20**, 367–387.

Frenkel, I. B. & Kac, V. G. 1980 Basic representations of affine Lie algebras and dual resonance models. *Invent. Math.* **62**, 23–66.

Frenkel, I. B., Lepowsky, J. & Meurman, A. 1988 *Vertex operator algebras and the monster* San Diego, California: Academic Press.

Friedan, D., Qiu, Z. & Shenker, S. 1984 Conformal invariance, unitarity and critical exponents in two dimensions. *Phys. Rev. Lett.* **52**, 1575–1578.

Goddard, P. & Olive, D. 1986 Kac–Moody and Virasoro algebras in relation to quantum physics. *Int. J. mod. Phys.* A **1**, 303–414.

Goddard, P. & Thorn, C. 1972 Comparability of the dual pomeron with unitarity and the absence of ghosts in the dual resonance model. *Phys. Lett.* B **103**, 207–210.

Goto, T. 1971 Relativistic quantum mechanics of one-dimensional mechanical continuum and subsidiary condition of dual resonance model. *Prog. theor. Phys.* **46**, 1560–1569.

Gliozzi, F., Scherk, J. & Olive, D. 1977 Supersymmetry, supergravity theories and the dual resonance models. *Nucl. Phys.* B **122**, 253–290.

Green, M. B. & Schwarz, J. H. 1981 Supersymmetrical dual string theory. *Nucl. Phys.* B **181**, 502–530.

Green, M. B. & Schwarz, J. H. 1984 Anomaly cancellations in supersymmetrical $D = 10$ gauge theory and superstring theory. *Phys. Lett.* B **149**, 117–122.

Gross, D. J., Harvey, J. A., Martinec, E. & Rohm, R. 1985 Heterotic string. *Phys. Rev. Lett.* **54**, 502–505.

Lovelace, C. 1971 Pomeron form factors and dual Regge cuts. *Phys. Lett.* B **34**, 500–506.

Neveu, A. & Schwarz, J. H. 1971 Factorisable dual models of pions. *Nucl. Phys.* B **31**, 86–112.

Olive, D. 1974 Dual models. In *Proc. XVIIth Int. Conf. High-Energy Physics* (ed. J. R. Smith), pp. 1269–1280.

Polyakov, A. M. 1981 Quantum geometry of bosonic strings. *Phys. Lett.* B **103**, 207–210.

Ramond, P. 1971 Dual theory for free fermions. *Phys. Rev.* D **3**, 2415-2418.

Segal, G. 1981 Unitary representations of some infinite dimensional groups. *Communs math. Phys.* **80**, 301–342.

Virasoro, M. 1970 Subsidiary conditions and ghosts in dual resonance models. *Phys. Rev.* D **1**, 2933–2936.

Discussion

P. T. LANDSBERG (*University of Southampton, U.K.*). In cosmology and elsewhere one is driven to talk about the 'coupling up' of the extra dimensions. Can Professor Olive comment on this problem.

D. I. OLIVE, F.R.S. I presume Professor Landsberg is referring to the 'curling up' of surplus dimensions of space whereby the conjugate momenta appear as internal charges. In string theory, as opposed to simpler theories, there is a more sophisticated and comprehensive version that I referred to when I mentioned the vertex operator construction for affine Kac–Moody algebras. I expect that other contributors, namely Goddard, Schwarz & Ross, have more to say about this.

Phil. Trans. R. Soc. Lond. A **329**, 329–342 (1989)

Printed in Great Britain

Gauge symmetry in string theory

By P. Goddard

*Department of Applied Mathematics and Theoretical Physics, University of Cambridge,
Silver Street, Cambridge CB3 9EW, U.K.*

An account is given of the way symmetry is incorporated into string theory by using the Frenkel–Kac–Segal mechanism, taken from the representation theory of affine Kac–Moody algebras. The intrinsically quantum mechanical nature of this mechanism is emphasized, and the present stage of development of string theory is compared with the 'old quantum theory'. The corresponding method of incorporating gauge symmetry into superstring theories is discussed and arguments that appear to prevent the construction of realistic theories of this type are reviewed in outline.

1. Introduction

The renaissance of string theory, which started in 1984 with the discovery of the anomaly cancellation by Green & Schwarz (1984), quickly led to the proposal of the heterotic string theory (Gross *et al.* 1985 *a*, *b*, *c*). This contained, in particular, two important conceptual developments. Each of these had been adumbrated in previous work but were now realized in a concrete, one might almost say blatant, fashion. Firstly, there was the realization that the left- and right-moving waves on a closed string can be treated to a large degree independently (though they are linked through considerations of modular invariance). This idea was there to some extent in the formulation of the Gliozzi–Olive–Scherk (gso) projection (Gliozzi *et al.* 1976, 1977), but in the heterotic string it reaches the extremity of having the left-moving modes vibrating in 26 dimensions in some sense, whereas the right-moving modes vibrate in 10, but have more degrees of freedom. (Because the extra dimensions for the left-moving modes are, in a sense we shall explain, intrinsically quantum mechanical, this perhaps somewhat disturbing picture need not be taken too literally.)

Secondly, and at least equally importantly, the heterotic string incorporated gauge symmetry in a new way, by using the Frenkel–Kac–Segal (fks) mechanism (Frenkel & Kac 1980; Segal 1981). Originally, internal symmetry was incorporated into string theory by using the Chan–Paton procedure (Paton & Chan 1969), which could be applied only to open string theories. It could be pictured as attaching to the ends of the string objects that had no inertia but merely carried the 'colour' quantum numbers of quarks (in the context of building the gauge symmetry of the strong interactions).

Although this procedure enabled Neveu & Scherk to show that non-abelian gauge symmetry could be incorporated inside string theory (Neveu & Scherk 1972), it was not wholly satisfactory for a number of reasons. It meant that closed strings were intrinsically neutral. Perhaps more fundamentally, it seems contrary to the general spirit of string theory to have the charges concentrated at two (or more) points of the string, rather than spread out along its length. The extra consistency claimed for string theories, as opposed to theories of point particles, is usually ascribed to the fact that strings are not all localized at points, and so it would seem somewhat at variance to the general philosophy of string theory to concentrate the charges in this way.

The FKS mechanism, which applies to closed string theories, describes the internal symmetry generators (and so the charges) as the integrals of local densities along the string. Thus the charges are, in a sense, smeared out. The work of Frenkel & Kac (1980), and of Segal (1981), was directed at constructing representations of affine Kac–Moody algebras using the vertex operators that had occurred in the development of dual models, before these models were reinterpreted as string theory. (In fact, such constructions had already occurred in special cases in the physics literature (Halpern 1975; Banks *et al.* 1976).) An attempt to elucidate the FKS mechanism in the context of the formalism commonly used in string theory, and to develop it further was made by Goddard & Olive (1984), who pointed out that this approach suggested that $\mathrm{Spin}(32)/\mathbb{Z}_2$ and $\mathrm{E}_8 \times \mathrm{E}_8$ might be particularly interesting possibilities for gauge groups. Of course, it was precisely for these gauge groups that Green & Schwarz subsequently found the anomaly cancellation, and for which the heterotic string theories could be constructed.

An important feature of the FKS mechanism, which has not always been stressed as much as it might in the literature, is that it is intrinsically quantum mechanical. We shall discuss this in some detail in §2 (see also Goddard 1987). This is to be contrasted with the Kaluza–Klein mechanism for internal symmetries, which incorporates the symmetry at the classical level. In this mechanism, space-time is enlarged from $\mathbb{R}^{3,1}$ to $\mathbb{R}^{3,1} \times \mathrm{M}$, where M is a compact manifold, and the internal symmetries are the isometries of M. In the FKS mechanism, we consider closed strings moving on a space of the form $\mathbb{R}^{3,1} \times \mathrm{T}$, a torus of dimension r, say. Classically, this gives rise to an internal symmetry of $\mathrm{U}(1)^r$, through a Kaluza–Klein type mechanism, but this symmetry is enhanced in the quantum theory to a non-abelian symmetry group G for a torus T of suitable dimensions, which are intrinsically quantum mechanical. The rank of G is the dimension r of the torus.

In the past four years there has been much frenetic activity in string theory, which has brought the subject to the attention of the wider scientific community, and even beyond, attracting much bemusement and a little derision. All this may be encouraging but it might well be argued that there has been much less conceptual progress in this period of intense activity since 1984 than in the first period of development, from 1968 to 1974 or 1976, when the subject was being pursued by a smaller band of physicists, whose efforts were less welcomed by the community at large. At that time there were major conceptual developments every year (e.g. Veneziano formula, n-point functions, higher-order contributions, Virasoro–Shapiro model, Virasoro conditions, fermionic models, supersymmetry, conformal field theory, no-ghost theorem, string picture, projection operators and the calculation of fermion scattering, connection with non-abelian gauge theory and gravity, GSO projection). In the intervening period, up to 1984, there were also a few major advances in our understanding of string theory, most notably Polyakov's (1981 *a, b*) path integral formulation (which made the connection with the methods of field theory much more immediate, thus making string theory more accessible and acceptable to a wider audience, and also facilitated the treatment of curved backgrounds, an essential step in substantiating the claims that string theory provides a successful quantum theory of gravity) and the proving of the space-time supersymmetry of the 10-dimensional theory, then renamed superstring theory (Green & Schwarz 1981).

Among the areas in which there has been conceptual development in the past four years, two related ones might be singled out (leaving aside string field theory, particularly Witten's (1986) beautiful approach, the ultimate status of which is unclear so far), namely modular invariance and conformal field theory. In the realization that ideas associated with the modular transformation properties of string partition functions could be useful in the classification of possible consistent string theories dates back to the work of Nahm (1976, 1977), which might have provided a greater stimulus had it not been published at a time when interest in string theory was rapidly diminishing. One of the recent advances in understanding has been to appreciate how crucial these properties are for the consistency of closed string theories, anomaly cancellation for example, and how they parallel considerations recently introduced in the study of the critical behaviour of two-dimensional statistical systems by Cardy (1986).

The mathematical language that unites two-dimensional critical phenomena with string theory is that of conformal field theory. In the context of string theory, the conformal field theory is the theory of the string degrees regarded as at theory defined over the two-dimensional world-sheet of the string. Elucidating the relation between the properties of the string in space-time and the properties of the conformal field theory defined by considering the dynamics of its degrees of freedom over its world-sheet, viewed as a two-dimensional 'world' itself, has been another area in which our understanding has increased recently. For example, a correspondence has been found between $N = 2$ superconformal invariance on the world-sheet and space-time supersymmetry (D. Friedan, A. Kent, S. Shenker & E. Witten, unpublished work 1987; Banks *et al.* 1988). In fact a programme has been outlined in which the conformal structure of string theory has put forward as the essential concept, with space and time being in some sense derived phenomenological concepts (Friedan & Shenker 1986) (though, a little paradoxically, this sophisticated approach might be regarded from a certain point of view as a step backwards, because realizing that the string world-sheet was to be thought of as embedded in space-time (Y. Nambu, unpublished work 1970; Goto 1971) was a very significant advance in the development of string theory). In any case it is widely expected that further advances in string theory will come from a progress in our understanding of conformal field theory.

In our review of some aspects of gauge symmetry in string theory, the ideas relating to conformal field theory and, although we shall not be able to discuss it in detail, modular invariance, will play an important role. Under the heading of conformal field theory may be subsumed considerations relating to the infinite-dimensional Lie algebras that enable us to understand much of the symmetry structure of string theories. (For a review of infinite-dimensional algebras in relation to quantum physics see Goddard & Olive (1986).) In §2 I shall discuss the FKS mechanism for the introduction of gauge symmetry into string theory, emphasizing its quantum-mechanical nature, and draw attention to some potentially paradoxical aspects of it. In §3 I discuss an analogous mechanism for gauge symmetry in superstring theories. This procedure enables me to obtain the observed gauge group, but not the correct representations for the quarks and leptons. I sketch the present obstacles to obtaining a realistic physical this way.

2. The Frenkel–Kac–Segal mechanism

The simplest geometric approach to string theory, and in many ways the one with the most intuitive physical picture, is that of Y. Nambu (unpublished work, 1970) and Goto (1971). This based on the action principle, directly generalizing the prescription that point particles move along geodesics, that classically the string moves in such a way as to extremize the area that it sweeps out in d-dimensional space-time (the area of its 'world-sheet'). Thus one has to find the extremum of the action

$$\mathscr{A} = \frac{T_0}{c} \int [(\dot{x}x')^2 - \dot{x}^2 x'^2]^{\frac{1}{2}} \, d\sigma \, d\tau, \tag{1}$$

where the world-sheet of the string is the two-dimensional surface $x^\mu(\sigma, \tau)$, $0 \leqslant \sigma \leqslant 2\pi$, $-\infty < \tau < \infty$, and the constant T_0 has been introduced to ensure that $\mathscr{A}$ has the correct dimensions of action; T_0 can be interpreted as the 'rest tension' in the string (Goddard *et al.* 1973). Here we shall only consider closed strings and so we have the condition $x(\sigma, \tau) = x(\sigma + 2\pi, \tau)$. The velocity of light has been denoted by c as usual and we use the notation

$$\dot{x} = \partial x / \partial \tau, \quad x' = \partial x / \partial \sigma. \tag{2}$$

If we quantize the simplest motions of the string, which are those in which it rotates, doubled up on itself into a straight line, in a plane, we see that the lowest quantum mechanical modes of the string have a characteristic length,

$$l = \sqrt{(c\hbar / \pi T_0)} \tag{3}$$

(which up to a constant could have been deduced on dimensional grounds). If the theory is accommodate gravity on a realistic scale, this length has to be comparable in magnitude with the Planck length

$$l_{\mathrm{P}} = \sqrt{(G\hbar / c^3)} \approx 10^{-33} \text{ cm}, \tag{4}$$

and then all the massive states of the string will have masses on the scale of the Planck mass $m_{\mathrm{P}} = \sqrt{(\hbar c / G)}$.

This leads to a potential paradox, for which there may be only partial explanations so far. The FKS mechanism, currently thought to be responsible for providing the observed symmetries of nature within the context of a string theory, requires that the string be moving in some of the 'internal' dimensions, i.e. those in excess of the familiar four, on a torus with a particular shape, as we shall discuss in greater detail in this section. The dimensions of this torus are fixed at values which are comparable with the Planck scale (4). Although the theory of a free string moving on such a toroidal background would be consistent in itself, we would not be content with this of course and we would wish to consider a theory of interacting strings. It is an inevitable consequence of the uncertainty principle in a quantum theory containing gravity that these fluctuations should be able to disturb the background. The fact that the fluctuations of the string have energies on a scale $m_{\mathrm{P}} c^2$ means that there will be fluctuations in the background geometry on a scale l_{P}, the scale of the torus itself. Given such large fluctuations in the background geometry responsible for it, one would expect the FKS symmetry to disappear, and we cannot consider a torus with dimension large on the Planck scale, because this would not possess the FKS symmetry.

One way round this potential difficulty is to suppose that the interaction between strings to be very weak, so that the shape providing the symmetry is disturbed very little. This would be a valid approach in string perturbation theory, sufficing perhaps to give a mathematical definition of string theory, but realistically we do not expect strings to be weakly coupled on the Planck scale in a realistic theory. Another way round this difficulty is to note that the observed symmetries in the heterotic string theory, for example, are supposed to come from the extra 16 dimensions available to the left-moving modes but not to the right-moving ones. This provides us with a rigid structure, incapable of deformation, like the affine Kac–Moody algebra resulting from it, and so unresponsive to the fluctuations we have been discussing. This is a potentially mathematically consistent answer to the 'paradox', but it might be thought not to be completely convincing physically, even if these extra 16 dimensions are not to be taken too seriously. It sounds rather like introducing a rigid wall into a physical problem, and this is procedure which, though superficially consistent, can often lead to inconsistencies on closer examination.

If we are not content with this explanation, we might regard the 'paradox' as rather reminiscent of that associated with the stability of the Bohr orbits of electrons in the old quantum theory. Bohr's original quantization conditions might be seen as analogous to rigidly fixing the dimensions of the torus in terms of multiples of l, the string length. Just as the Bohr orbits could be destroyed, within the pseudo-classical framework of their formulation, by allowing the electron to radiate (thus removing the explanation of the observed discrete spectral lines, which was their *raison d'être*) so the torus can be destroyed by allowing the strings to interact with their background. If there is a real difficulty here, it may be that its resolution will require a conceptual revolution, associated with understanding physical phenomena on length scales of the order of l_P, just as quantum theory required a revolution associated with the description of phenomena with action comparable with $\hbar$. Because string theory has introduced a third, and *presumably* the last, dimension-bearing constant, l (after c and $\hbar$), it would be surprising, and perhaps disappointing, if it did not require a complete revision of our concepts. These conceptual revolutions (relativity and quantum mechanics) have been successively more drastic and have taken us further from the world of common experience. They have been less immediately accessible. This should be even more so in this third case. A major aspect of such a revolution may well be the abolition of the space-time continuum, and there have been various suggestions along these lines in the context of string field theory (Witten 1986) and conformal field theory (Friedan & Shenker 1986), but it may be that something very much more radical, making a departure from quantum theory as well, is required. A more recent extremely interesting suggestion for the nature of a theory on length scales less than l_P, both radical and plausible, is provided by Witten (1988).

As a preliminary to describing the FKS mechanism, we shall now review in outline the quantization of the string (Goddard *et al.* 1973). Choosing orthonormal coordinates σ, τ such that

$$\dot{x}^2 + x'^2 = 0, \quad x'\dot{x} = 0, \tag{5}$$

the equation of motion of the string, which follows from extremizing the action (1), is

$$\ddot{x} = x''. \tag{6}$$

The momentum conjugate to $x^\mu(\sigma, \tau)$ is

$$\pi^\mu(\sigma, \tau) = (T_0/c)\,\dot{x}^\mu(\sigma, \tau), \tag{7}$$

so that the canonical commutation relations are

$$[x^\mu(\sigma, \tau), \pi^\nu(\sigma', \tau)] = i\hbar\delta(\sigma - \sigma')\,\eta^{\mu\nu}, \tag{8}$$

where $\eta^{\mu\nu}$ denotes the space-time metric (taken to be $\eta^{00} = -1$; $\eta^{jj} = 1$ for each j, $1 \leqslant j \leqslant d-1$; $\eta^{\mu\nu} = 0$, $\mu \neq \nu$).

Consider first a closed string moving in $\mathbb{R}^{25,1}$. (For the bosonic string theory that we are considering, the space has to be 26-dimensional for consistency of the theory.) Then, the general solution to (6) can be written

$$x(\sigma,\tau)/l = \tfrac{1}{2}X_{\mathrm{L}}(\mathrm{e}^{i(\tau+\sigma)}) + \tfrac{1}{2}X_{\mathrm{R}}(\mathrm{e}^{i(\tau-\sigma)}),\tag{9}$$

where

$$X_{\mathrm{L}}(z) = q - i p_{\mathrm{L}} \ln z + i \sum_{n\neq 0} \frac{\alpha_n^{\mathrm{L}}}{n} z^{-n}\tag{10a}$$

and

$$X_{\mathrm{R}}(z) = q - i p_{\mathrm{R}} \ln z + i \sum_{n\neq 0} \frac{\alpha_n^{\mathrm{R}}}{n} z^{-n},\tag{10b}$$

and the factor of l^{-1} has been included to leave q, p and α dimensionless.

The periodicity boundary condition $x(\sigma,\tau) = x(\sigma+2\pi)$, applied to all the components of x, then implies

$$p_{\mathrm{L}} = p_{\mathrm{R}} = p,\tag{11}$$

say. The canonical commutation relations (8) are then equivalent to

$$[\alpha_m^{\mathrm{R}\mu}, \alpha_n^{\mathrm{R}\nu}] = [\alpha_m^{\mathrm{L}\mu}, \alpha_n^{\mathrm{L}\mu}] = m\delta_{m,-n}\,\eta^{\mu\nu}, \quad [\alpha_m^{\mathrm{L}\mu}, \alpha_n^{\mathrm{R}\mu}] = 0,\tag{12}$$

$$[q^\mu, p^\nu] = \tfrac{1}{2}i\eta^{\mu\nu}.\tag{13}$$

This means that p^μ has the representation

$$p_\mu = \tfrac{1}{2}i\,\partial/\partial q^\mu.\tag{14}$$

We consider the case in which the whole space is not $\mathbb{R}^{25,1}$ but some of the directions are compact. Consider the simplest case in which one of the directions, x^1, is compactified into a circle of radius $R = la$ (so that a is a dimensionless parameter). Because the string might wind round the circle some integral number of times, m, say, the periodicity condition for the x^1 component is replaced by

$$x^1(\sigma,\tau) = x^1(\sigma+2\pi,\tau) + 2\pi m R.\tag{15}$$

Note that the integer m, the winding number is a *classical* concept. The condition (15) implies that, rather than (11), $p_{\mathrm{L}}^1, p_{\mathrm{R}}^1$ satisfy

$$p_{\mathrm{L}}^1 - p_{\mathrm{R}}^1 = 2ma,\tag{16}$$

and the corresponding quantum conditions are

$$[q^1, p_{\mathrm{L}}^1] = [q^1, p_{\mathrm{R}}^1] = \tfrac{1}{2}i,\tag{17}$$

which have the representation

$$p_{\mathrm{L}}^1 = \tfrac{1}{2}i\,\partial/\partial q^1 + ma, \quad p_{\mathrm{R}}^1 = \tfrac{1}{2}i\,\partial/\partial q^1 - ma.\tag{18}$$

Now the wave function of the string has to be single valued on the circle and so its dependence on q^1 must be a sum of terms of the form

$$\mathrm{e}^{inq^1/a}\tag{19}$$

for some integer n. This is the familiar quantization of momentum resulting from the periodicity on a circle. The quantum number n, in contrast to the winding number m, is indeed a *quantum* concept.

[16]

If we consider the two parts of the momentum, left-moving and right-moving, as part of a single vector we can write

$$(p_{\rm L}^1; p_{\rm R}^1) = (n/2a + ma; n/2a - ma). \tag{20}$$

Because m and n, for apparently very different reasons, are integers, we see that the vector (20) lies on a lattice in $\mathbb{R}^2$. If we introduce an indefinite inner product on this two-dimensional real space, so regarding it as $\mathbb{R}^{1,1}$

$$(p_{\rm L}^1; p_{\rm R}^1)^2 \equiv (p_{\rm L}^1)^2 - (p_{\rm R}^1)^2 = 2mn, \tag{21}$$

we see that the inner product of any two points of this momentum lattice is an even integer. A lattice with this property is an *even* lorentzian lattice. In fact this lattice is also *self-dual*, that is it is the same as its dual or reciprocal lattice, the lattice consisting of all points having integral inner products with all points of the lattice defined by (20). Such even self-dual lattices can only exist in spaces $\mathbb{R}^{M,N}$ for which $M - N$ is a multiple of 8. Their particular significance in relation to string theory is one of the realizations of the past few years (Frenkel 1985; Goddard & Olive 1984; Narain 1986; for a recent review see Lerche *et al.* 1989).

We note that this lattice is invariant under the replacement

$$a \to 1/2a. \tag{22}$$

This operation interchanges the classical winding number m with the quantum number n. Such a symmetry is very analogous to the symmetry (Goddard *et al.* 1977) between the quantum-mechanical electric charges and the classical magnetic charges that exist in spontaneously broken gauge theories that possess classical solutions, of the 't Hooft–Polyakov type, bearing magnetic charge. The classical and quantum nature of m and n is made clear by writing down the corresponding component of the total left-moving momentum of the string

$$\mathscr{P}_{\rm L}^1 = 2\pi T_0\, l p_{\rm L}^1/c = m R T_0/c + n\hbar/2R, \tag{23}$$

the first term on the right-hand side being classical and the second quantum. In terms of the original radius R, the replacement (22) is

$$R \to c\hbar/\pi T_0 R, \tag{24}$$

making clear the quantum-mechanical nature of the transformation.

The FKS symmetry occurs at the fixed point of this transformation, namely at $a = \frac{1}{\sqrt{2}}$. At this point the symmetry of the quantum theory of the closed string increases from U(1) to SU(2) (or, more precisely, from U(1) × U(1) to SU(2) × SU(2)). Thus the symmetry is indeed intrinsically quantum mechanical.

A signal of the increased gauge symmetry appearing at $a = \frac{1}{\sqrt{2}}$ is the appearance of extra massless states. Suppose we denote states of the string with momentum but no excitations by

$$|k; k_{\rm L}^1; k_{\rm R}^1\rangle, \tag{25}$$

where k denotes the momentum, both left- and right-moving, in the non-compact directions and $k_{\rm L}^1$ and $k_{\rm R}^1$ denote the components in the compact direction. Then they are always massless states of the string

$$\alpha_{-1}^{{\rm L}1} \alpha_{-1}^{{\rm R}\mu} |k; 0; 0\rangle, \tag{26}$$

where $\mu \neq 1$ and $k^2 = 0$. (We are using 'mass' in the usual sense of being associated with the non-compact dimensions of the space-time in which the string is moving.) These states

[17]

336   P. GODDARD

correspond to a massless vector particle, the gauge particle associated with the U(1) symmetry. (There is a similar gauge particle, obtained by interchanging left and right, associated with the other U(1) symmetry.) At $a = \frac{1}{\sqrt{2}}$, extra massless states appear:

$$\alpha_{-1}^{R\mu} |k; \pm \sqrt{2}; 0\rangle, \tag{27}$$

where again $\mu \neq 1$ and $k^2 = 0$. These are the extra gauge bosons corresponding to the enhancement of the U(1) symmetry to SU(2) through the FKS mechanism.

The states (26) and (27) satisfy the constraints required of physical states of the closed string in this covariant formalism

$$L_n^L |\psi\rangle = 0, \quad L_n^R |\psi\rangle = 0, \quad n > 0, \tag{28a}$$

$$(L_0^L + L_0^R) |\psi\rangle = 2 |\psi\rangle, \tag{28b}$$

$$L_0^L |\psi\rangle = L_0^R |\psi\rangle, \tag{29}$$

where L_n^L is defined in terms of α^L and $\alpha_0^L = p^L$, and L_n^R is defined in terms of α^R and $\alpha_0^R = p^R$, by equations of the form:

$$L(z) \equiv \sum_{n=-\infty}^{\infty} L_n z^{-n-2} = \frac{1}{2} : \left(i\frac{dX}{dz}\right)^2 :, \tag{30}$$

$$L_n = \frac{1}{2} \sum_{m=-\infty}^{\infty} :\alpha_m \alpha_{n-m}:. \tag{31}$$

The colons denote normal ordering with respect to the operators α_m. The L_n defined by these equations satisfy the Virasoro algebra

$$[L_m, L_n] = (m-n) L_{m+n} + \tfrac{1}{12} cm(m^2 - 1) \delta_{m,-n}, \tag{32}$$

with the *central charge* $c = 26$, the dimension of space-time required for consistency.

The considerations we have discussed here generalize from compactification on a circle to compactification on some more general torus. We can construct such a torus by identifying points in space related by displacements lying on some lattice Λ, whose points span the dimensions that are being compactified. This means that we view space-time as a quotient $\mathbb{R}^{25,1}/\Lambda$. In the simple example of compactification on a circle, we have $\Lambda = 2\pi R\mathbb{Z}$. Generalizing $R = la$, it is convenient to write $\Lambda = 2\pi l \Gamma$. Then, instead of (16) we have

$$p_L - p_R = 2\gamma \quad \text{for some} \quad \gamma \in \Gamma. \tag{33}$$

For the string's wave function to be single valued, it has to be a sum of terms whose dependence on q is of the form $e^{i\beta q}$, where β has integral inner product with all the points of the lattice Γ. This means that the projection β' of β onto the space spanned by Γ has to lie on the lattice Γ^* dual to Γ. It then follows that the corresponding projections p_L', p_R', of p_L and p_R, respectively, can be put together to form the vector

$$(p_L'; p_R') = (\tfrac{1}{2}\beta' + \gamma, \tfrac{1}{2}\beta' - \gamma), \quad \gamma \in \Gamma, \quad \beta' \in \Gamma^*. \tag{34}$$

This again forms an even self-dual lattice, in $\mathbb{R}^{M,M}$, where $M = \dim \Gamma$ (Englert & Neveu 1985; Narain 1986).

As we saw in the case of compactification on the circle, for particular quantum-mechanical values of the dimensions of such a torus, the extra massless particles appear, indicating enhancement of symmetry. The generators of such symmetries can be constructed from the

[18]

vertex operators describing the emission of these massless vector particles. For definiteness, consider massless particles whose momentum in the compact dimensions is entirely associated with the left-moving modes on the string and whose momentum in the uncompactified dimensions is associated with the right-moving modes. The vertex describing the emission of such a particle will consist of the product of a right-moving part and a left-moving part. The right-moving part is of the form,

$$i\,dX_{\rm R}^{\mu}/dz : e^{ikX_{\rm R}} : . \tag{35}$$

If we denote the left-moving part of the vertex by $T^a(z)$, under quite general considerations the moments of $T^a(z)$ will satisfy an affine Kac–Moody algebra, $\hat{g}$,

$$[T_m^a, T_n^b] = if^{abc} T_{m+n}^c + km\delta_{m,-n}\delta^{ab}, \tag{36}$$

where

$$T^a(z) = \sum_{n=-\infty}^{\infty} T_n^a z^{-n-1}. \tag{37}$$

The moments will also satisfy

$$[L_m, T_n^a] = -nT_{m+n}^a. \tag{38}$$

The zero moments T_0^a satisfy the Lie algebra of a compact group, with Lie algebra g,

$$[T_0^a, T_0^b] = if^{abc} T_0^c, \tag{39}$$

which is the gauge group provided by the FKS mechanism. The quantity k determines the *level* of the affine Kac–Moody algebra. (For further explanation of the FKS mechanism see, for example, Green *et al.* (1987) or Goddard (1987); for a review of Kac–Moody algebras in relation to their applications in physics see Goddard & Olive (1986).)

3. Type II strings

As we have remarked, to get consistency in the bosonic string theory discussed in the last section we need $c = 26$ in (32). To get a theory in four-dimensional space-time, we need to compactify 22 of the dimensions. Even compactifying these, on a suitable torus for example, we would still be left with a tachyonic state (a state of negative squared mass) as a bar to any sort of realistic physical interpretation. At present there are basically two sorts of ways of avoiding tachyons. Firstly, there are the superstring theories, in which fermionic degrees of freedom are added to the bosonic. This reduces the 'critical dimension', in which the theory achieves consistency, from 26 to 10, and the value of c in (32) is now 15, receiving a contribution of one from each bosonic degree of freedom and a half from each fermionic one. We restrict attention here to the closed superstring theories (type II theories), considering the introduction of symmetry by a fermionic analogue of the FKS procedure. Secondly, there are the heterotic string theories, which, as we mentioned in §1, are a sort of hybrid between the superstring and the bosonic string. The gauge symmetry here comes from compactification of the extra bosonic left-moving degrees of freedom, using the FKS mechanism, and so producing, before any symmetry breaking, a rank sixteen gauge group, $E_8 \times E_8$ or $Spin(32)/\mathbb{Z}_2$.

The large extra number of dimensions in the heterotic string theory results in a much larger gauge symmetry than has so far been observed. After compactification of the 16 extra bosonic degrees of freedom, one is left with a 10-dimensional theory. It is still necessary to compactify a further six to obtain a four-dimensional world. This further compactification has been typically used to break the large symmetry resulting from the FKS mechanism to something more compatible with observation. However, if it were done in a symmetrical way, and it might be that in such a further compactification one is driven to the sort of fixed of a transformation like (22), the symmetry would be further enhanced rather than reduced.

We shall consider here attempts that have taken place in the last two years to incorporate gauge symmetry into type II superstring theories (Bluhm *et al.* 1987, 1988; Kawai *et al.* 1987; Antoniadis *et al.* 1987). These attempts have succeeded in showing that, contrary to previous expectations, it is possible to incorporate non-abelian symmetries into such theories, including ones large enough to contain the gauge group, $SU(3) \times SU(2) \times U(1)$, of the 'standard model'. At the moment there seems unfortunately to be a firm obstacle to obtaining a realistic spectrum of quarks and leptons (Dixon *et al.* 1987).

In the superstring, in addition to the bosonic degrees of freedom moving each way, described by $X''(z)$, there are fermionic degrees of freedom described by fermion fields, $\psi^\mu(z)$, $1 \leqslant \mu \leqslant d$, where d is the dimension of space-time. Now the expression for the Virasoro generators has to include both bosonic and fermionic degrees of freedom,

$$L(z) = \frac{1}{2} : \left(i \frac{\mathrm{d}X}{\mathrm{d}z} \right)^2 : + \frac{1}{2} : \frac{\mathrm{d}\psi}{\mathrm{d}z} \psi :, \tag{40}$$

and the algebra defining the physical state conditions is extended to include 'super-Virasoro' or 'superconformal' generators,

$$G(z) \equiv \sum_r G_r z^{-r-\frac{3}{2}} = \psi(z) \, i \frac{\mathrm{d}X}{\mathrm{d}z}. \tag{41}$$

The final algebra is now

$$[L_m, L_n] = (m-n) L_{m+n} + \tfrac{1}{12} cm(m^2-1) \, \delta_{m,-n}, \tag{42a}$$

$$[L_m, G_r] = (\tfrac{1}{2}m - r) G_{m+r}, \tag{42b}$$

$$\{G_r, G_s\} = 2L_{r+s} + \tfrac{1}{3}c(r^2 - \tfrac{1}{4}) \, \delta_{r,-s}. \tag{42c}$$

Here, m, n are integers and r, s can be either half-odd-integers (Neveu–Schwarz case) or integers (Ramond case), the former corresponding to excitations being space-time bosons and the latter to their being space-time fermions. In (42), $c = 15$ for consistency, corresponding to a space-time dimension of 10, with a contribution of 1 from each boson and $\tfrac{1}{2}$ from each fermion. The fermion field has an expansion of the form

$$\psi(z) = \sum_r \psi_r z^{-r-\frac{1}{2}}. \tag{43}$$

The physical state conditions take the form

$$L_n |\psi\rangle = 0, \quad G_r |\psi\rangle = 0, \quad n, r > 0, \tag{44a}$$

$$(L_0 - \tfrac{1}{2}) |\psi\rangle = 0. \tag{44b}$$

In the case of the type II superstring, there is an algebra and set of conditions like this for both the left- and the right-moving modes. (For the heterotic string, the left-moving modes have only bosonic degrees of freedom, carrying the internal symmetry, $c = d = 26$, whereas the right-moving modes have $d = 10$, $c = 15$.) To get a four-dimensional theory, one must compactify six of the 10 dimensions, regarding them as internal.

Initially, it was thought that type II strings could only have $U(1)$ gauge symmetry. Then it was realized that for suitable correlations of left- and right-moving modes, at the critical radius, one could get $SU(2)^6$ symmetry (Bluhm & Dolan 1986), which was non-abelian but not very interesting phenomenologically. Finally, it was seen how this could be generalized to the groups $SU(2) \times SU(4)$ and $SO(5) \times SU(3)$, or subgroups thereof. To describe this procedure for incorporating gauge symmetry into the superstring, it is best to replace the six internal boson degrees of freedom by 12 internal fermions, making 18 internal fermions in all. Then the part of (40) corresponding to the six internal dimensions can be rewritten

$$L_{\text{int}}(z) = \tfrac{1}{2} : \psi^a \, d\psi^a / dz :, \tag{45}$$

where the sum is over $a = 1$ to 16. The method of introducing gauge symmetry is to change the internal part of (40) to
$$G_{\text{int}}(z) = -\tfrac{1}{6} i f_{abc} \, \psi^a(z) \, \psi^b(z) \, \psi^c(z), \tag{46}$$

where f_{abc} are the structure constants of an 18-dimensional semisimple Lie algebra, g, which gives the possibilities $SU(2)^6$, $SU(2) \times SU(4)$ and $SO(5) \times SU(3)$ (Bluhm $et\ al.$ 1987). It can be shown (Goddard & Olive 1985) that (45) and (46) satisfy a super-Virasoro algebra with a central charge $c_{\text{int}} = 9$, and the theory can be arranged so that it has a corresponding gauge symmetry whose generators T_0^a are moments of

$$T^a(z) = -\tfrac{1}{2} i f_{abc} \, \psi^b(z) \, \psi^c(z). \tag{47}$$

This symmetry can easily be broken from g to $g' \subset g$, where g/g' is a symmetric space. It is in fact possible to obtain a type II theory with a gauge symmetry group which is any subgroup of $SU(2)^6$, $SU(4) \times SU(2)$, $SO(5) \times SU(3)$, $SU(3) \times SU(2)^2$, $SU(3)^2$ or G_2 (Dixon $et\ al.$ 1987).

In many ways these models appear very attractive. It is possible to obtain representation contents which are tantalizingly close to those observed in nature (Bluhm $et\ al.$ 1988). The states one focuses on are those that are massless. According to the currently accepted interpretation of string theory, it is only these states that should be directly physically observable in practice, because all other states will have masses of the order of m_{P}, far above accessible energies. The massless states will have to acquire their masses as the result of some symmetry breaking, often ascribed to some conjectured non-perturbative effect. It seems not too difficult to construct models that seem fairly natural, and in which the observed states and few others occur, but as the massless level and the first excited level rather than all at the massless level. It is difficult to know what significance to attach to these 'near misses'. In fact there is the rather general argument of Dixon $et\ al.$ (1987) to the effect that it is impossible to construct a type II model with a gauge group containing that of the standard model, $SU(3) \times SU(2) \times U(1)$, with massless states transforming non-trivially under both $SU(3)$ and $SU(2)$, as we need for certain of the quark states in nature.

As a preliminary to sketching this argument, note that if we have an affine algebra as in (36) and (38), we can place a lower bound on the value of the central charge of the Virasoro algebra. To do this we make use of the Sugawara construction of a Virasoro algebra out of an affine Kac–Moody algebra, defining

$$\mathscr{L}_n^g = \frac{1}{2k + Q^g} \sum_m {}^\times_\times T_m^a \, T_{n-m}^a {}^\times_\times, \tag{48}$$

where the normal ordering operation denoted by the crosses is defined by

$$\left. \begin{aligned} {}^\times_\times T_m^a \, T_n^{a\,\times}_{\times} &= T_m^a \, T_n^a, \quad \text{if } n \geqslant 0, \\ &= T_n^a \, T_m^a, \quad \text{if } n \leqslant 0, \end{aligned} \right\} \tag{49}$$

[21]

the quantity Q^g is the quadratic Casimir operator of g in the adjoint representation,

$$f^{abc}f^{abd} = Q^g\delta^{cd}. \tag{50}$$

The $\mathscr{L}^g$ satisfy the Virasoro algebra with central charge

$$c^g = 2k\dim g/(2k+Q^g) \tag{51a}$$

$$= x\dim g/(x+h^g). \tag{51b}$$

The number $x = 2k/\Psi^2$, where Ψ is a long root of the Lie algebra g, is called the *level* of the representation of $\hat{g}$; $h^g = Q^g/\Psi^2$ is called the dual Coxeter number of g. Both x and h^g have to be integers. If we consider

$$K_n = L_n - \mathscr{L}^g_n, \tag{52}$$

we obtain a Virasoro algebra with a value of the central charge $c^K = c - c^g$. Because the representation of the Virasoro algebra K is easily seen to be unitary, at least if we restrict attention to L_{int}, and with the spectrum of K_0 bounded below, it follows that $c^K \geqslant 0$ and so

$$c_{\text{int}} \geqslant c^g. \tag{53}$$

This can be viewed as an example of the coset construction (Goddard *et al.* 1985, 1986).

In the case of type II theories, the gauge particles could come from either both the Neveu–Schwarz sectors of both the left- and right-moving modes, or the Ramond sector of both modes. (The particles that come from the Neveu–Schwarz sector of one and the Ramond sector of the other are space-time fermions.) Assuming that they come from the Neveu–Schwarz sector of both, let us consider the gauge symmetry resulting from the construction (46) applied to the left-moving modes. There are in superstring theories two parts to the vertex describing such a particle, which in this case consists of a bosonic part, $T^a(z)$, and a fermionic part, $\chi^a(z)$. One can show on quite general grounds that the moments of these satisfy

$$[T^a_m, T^b_n] = \mathrm{i}f^{abc}T^c_{m+n} + km\delta_{m,-n}\delta^{ab}, \tag{54a}$$

$$[T^a_m, \chi^b_r] = \mathrm{i}f^{abc}\chi^c_{m+r}, \tag{54b}$$

$$\{\chi^a_r, \chi^b_s\} = \delta_{r,s}\delta^{a,b}, \tag{54c}$$

a super-Kac–Moody algebra. Given such a construction, we can perform a subtraction (rather as in the coset construction), and set

$$\tilde{T}^a(z) = T^a(z) - \tfrac{1}{2}\mathrm{i}f^{abc}\chi^b(z)\,\chi^c(z). \tag{55}$$

Then $\tilde{T}^a(z)$ will provide a representation of the affine Kac–Moody algebra (54a) which commutes with $\chi^b(z)$. We can then construct two commuting Virasoro algebras, the Sugawara construction applied to $\tilde{T}^a(z)$, $\tilde{\mathscr{L}}^g(z)$, and the Virasoro algebra, $L^\chi(z)$, associated with the free fermion fields $\chi^a(z)$. Again

$$K(z) = L_{\text{int}}(z) - \tilde{\mathscr{L}}^g(z) - L^\chi(z) \tag{56}$$

is a Virasoro algebra, commuting with both $T^a(z)$ and $\chi^b(z)$, and with central charge $c^K = c_{\text{int}} - \tilde{c}^g - \tfrac{1}{2}\dim g$, where $\tilde{c}$ is the central charge for $\tilde{\mathscr{L}}^g$ and $\tfrac{1}{2}\dim g$ is that for L^χ. Thus it follows that

$$c_{\text{int}} \geqslant \tilde{c} + \tfrac{1}{2}\dim g. \tag{57}$$

[22]

We are interested in the case $g = su(3) \oplus su(2) \oplus u(1)$, for which $\dim g = 12$, and

$$\tilde{c}^g = \tilde{c}^{su(3)} + \tilde{c}^{su(2)} + \tilde{c}^{u(1)}, \tag{58}$$

so that
$$c_{\text{int}} \geqslant 10, \tag{59}$$

whereas we should have $c_{\text{int}} = 9$ for a type II theory, as there are six internal dimensions. This is the basic contradiction, or obstacle to getting a realistic type II string theory at present.

There are some further points to be made. If the non-abelian gauge symmetry is associated with the left-moving modes, it can be shown that the fermions for which the left-moving modes are of Ramond type cannot be massless (and so correspond to particles that might in practice be observable). Hence, the non-abelian gauge symmetry cannot be divided between the left- and right-moving modes. Moreover, if there is a right-moving $U(1)$ gauge symmetry, the massless fermions can be shown to be neutral under it, so in practice it is of little relevance. The only remaining obvious possibility is that some or all of the gauge particles are of Ramond–Ramond type. Dixon *et al.* also present argument to exclude this, though it is perhaps in this region that the best hope for a chink in the argument lies.

All this said, it is intriguing that just one unit of c, just a '10 % effect', bars us from producing a more nearly realistic type II string theory. The progress of string theory, and our understanding of its structure and interpretation, has been punctuated by 'no-go theorems' of this sort. Usually, the way forward has been to the side (which could perhaps point to heterotic string theory), or to take more seriously what string theory itself is insisting on as its own interpretation. Given that the philosophy of string theory is that, once all the consistency conditions are known, the correct physical theory should be uniquely determined, some reason has in any case to be found why these type II theories are inconsistent, not merely phenomenologically unacceptable. In any case, the delicacy of the discrepancy might suggest that the mechanisms described in this section deserve further consideration.

REFERENCES

Antoniadis, I., Bachas, C. & Kounnas, C. 1987 *Nucl. Phys.* B **289**, 87.
Antoniadis, I. & Bachas, C. 1988 *Nucl. Phys.* B **298**, 586.
Banks, T., Dixon, L. J., Friedan, D. & Martinec, E. 1988 *Nucl. Phys.* B **299**, 613.
Banks, T., Horn, D. & Neuberger, H. 1976 *Nucl. Phys.* B **108**, 119.
Bluhm, R. & Dolan, L. 1986 *Phys. Lett.* B **169**, 347.
Bluhm, R., Dolan, L. & Goddard, P. 1987 *Nucl. Phys.* B **289**, 364.
Bluhm, R., Dolan, L. & Goddard, P. 1988 *Nucl. Phys.* B **309**, 330.
Cardy, J. L. 1986 *Nucl. Phys.* B **270** [**FS18**], 445.
Dixon, L., Kaplunovsky, V. & Vafa, C. 1987 *Nucl. Phys.* B **294**, 43.
Englert, F. & Neveu, A. 1985 *Phys. Lett.* B **163**, 349.
Frenkel, I. B. 1985 *Lectures in applied mathematics*, vol. 21, p. 325. American Mathematical Society.
Frenkel, I. B. & Kac, V. G. 1980 *Invent. Math.* **62**, 23.
Friedan, D. & Shenker, S. 1986 *Phys. Lett.* B **175**, 287.
Gliozzi, F., Scherk, J. & Olive, D. 1976 *Phys. Lett.* B **65**, 282.
Gliozzi, F., Scherk, J. & Olive, D. 1977 *Nucl. Phys.* B **122**, 253.
Goddard, P. 1987 In *Superstrings '87* (ed. L. Alvarez-Gaumé *et al.*), pp. 157–189. Singapore: World Scientific.
Goddard, P., Goldstone, J., Rebbi, C. & Thorn, C. B. 1973 *Nucl. Phys.* B **56**, 109.
Goddard, P., Kent, A. & Olive, D. 1985 *Phys. Lett.* B **152**, 88.
Goddard, P., Kent, A. & Olive, D. 1986 *Communs math. Phys.* **103**, 105.
Goddard, P., Nuyts, J. & Olive, D. 1977 *Nucl. Phys.* B **125**, 1.
Goddard, P. & Olive, D. 1984 In *Vertex operators in mathematics and physics*, *Proc. Conf.*, November 10–17, 1983 (ed. J. Lepowsky, S. Mandelstam & I. M. Singer), pp. 51–96. New York: Springer-Verlag.
Goddard, P. & Olive, D. 1985 *Nucl. Phys.* B **257** [**FS14**], 226.

Goddard, P. & Olive, D. 1986 *Int. J. mod. Phys.* A **1**, 303.

Goto, T. 1971 *Prog. theor. Phys.* **46**, 1560.

Green, M. B. & Schwarz, J. H. 1981 *Nucl. Phys.* B **181**, 502.

Green, M. B. & Schwarz, J. H. 1984 *Phys. Lett.* B **149**, 117.

Green, M. B., Schwarz, J. H. & Witten, E. 1987 *Superstring theory*, vol. 1 (*Introduction*). Cambridge University Press.

Gross, D. J., Harvey, J. A., Martinec, E. & Rohm, R. 1985a *Phys. Rev. Lett.* **54**, 502.

Gross, D. J., Harvey, J. A., Martinec, E. & Rohm, R. 1985b *Nucl. Phys.* B **256**, 253.

Gross, D. J., Harvey, J. A., Martinec, E. & Rohm, R. 1985c *Nucl. Phys.* B **267**, 75.

Halpern, M. 1975 *Phys. Rev.* D **4**, 1684.

Kawai, H., Lewellen, D. C. & Tye, S.-H. H. 1987 *Phys. Lett.* B **191**, 63.

Lerche, W., Schellekens, A. N. & Warner, N. P. 1989 *Phys. Rep.* (In the press.)

Nahm, W. 1976 *Nucl. Phys.* B **114**, 174.

Nahm, W. 1977 *Nucl. Phys.* B **120**, 125.

Narain, K. S. 1986 *Phys. Lett.* B **169**, 41.

Neveu, A. & Scherk, J. 1972 *Nucl. Phys.* B **36**, 35.

Paton, J. E. & Chan, H. M. 1969 *Nucl. Phys.* B **10**, 516.

Polyakov, A. M. 1981a *Phys. Lett.* B **103**, 207.

Polyakov, A. M. 1981b *Phys. Lett.* B **103**, 211.

Segal, G. 1981 *Communs math. Phys.* **80**, 301.

Witten, E. 1986 *Nucl. Phys.* B **268**, 23.

Witten, E. 1988 *Communs math. Phys.* **117**, 353.

Quantum groups and conformal field theories

By L. Alvarez-Gaumé

European Laboratory for Particle Physics, European Organization for Nuclear Research,
CH-1211 Geneva, Switzerland

Rational conformal field theories can be interpreted as defining quasi-triangular Hopf algebras. The Hopf algebra is determined by the duality properties of the conformal theory.

Important advances have been made recently towards the classification of rational conformal field theories (RCFT). An RCFT is characterized by a chiral algebra $\sigma = \sigma_L \times \sigma_R$ such that σ_L (σ_R) contains at least the identity operator and the Virasoro algebra, and the Hilbert space H of the theory splits into a finite number of irreducible representations of $\sigma : H = \oplus H_i \times H_{\bar{\imath}}$, with $i, \bar{\imath}$ running over a finite range of values. Examples are provided by the minimal models of Belavin *et al.* (1988) and the discrete unitary series of Virasoro representations (Friedan *et al.* 1984) whose chiral algebra is the Virasoro algebra; the two-dimensional Wess–Zumino–Witten theory (Witten 1984) with σ_L an affine Kac–Moody algebra, etc. A classification of RCFTs is important in the determination of universality classes of two-dimensional critical systems and it may also be an important step towards the resolution of the far more difficult problem of understanding the space of classical solutions to string theories.

Verlinde (1988) studied the fusion algebra of an RCFT, which is a consequence of the operator algebra of the theory. The structure constants of this algebra are given by the different couplings between three conformal families. If $[\phi_i]$ denotes the conformal family of the primary field ϕ_i, the fusion algebra is written as

$$[\phi_i] \times [\phi_j] = \sum_k N_{ij}{}^k [\phi_k] \tag{1}$$

and the $N_{ij}{}^k$ are non-negative integers. If we define the matrices $(N_i)_j{}^k = N_{ij}{}^k$, the associativity of the operator algebra of the conformal theory implies that the N_is commute. More abstractly, the fusion algebra is a commutative associative algebra with as many generators as conformal families in the theory and with structure constants $N_{ij}{}^k$. For each family $[\phi_i]$ we can construct its character:

$$\chi_i(\tau) = \mathrm{Tr}_{[\phi_i]}\, q^{L_0 - \frac{1}{24}c}, \quad q = e^{2\pi i \tau}. \tag{2}$$

The behaviour of (2) under modular transformations in a modular covariant theory is:

$$\left. \begin{aligned} T: \quad & \chi_i(\tau+1) = e^{2\pi i(h_i - \frac{1}{24}c)} \chi_i(\tau), \\ S: \quad & \chi_i(-1/\tau) = S_i^j \chi_j(\tau), \end{aligned} \right\} \tag{3}$$

where h_i is the conformal dimension of ϕ_i, and c is the central extension of the Virasoro algebra. Verlinde (1988) showed by many examples that the matrix S diagonalizes the fusion rules. More precisely, if we write $N_{ij}{}^0 = C_{ij}$ and use C to lower indices, then

$$N_{ijk} = \sum_m \frac{S_{im} S_{jm} S_{km}}{S_{0m}} \tag{4}$$

[25]

and the eigenvalues of N_i are $\lambda_i^{(k)} = S_i^{\ k}/S_0^{\ k}$. This striking connection between modular transformations and the fusion rules was proved rigorously by Moore & Seiberg (1989) using a set of polynomial equations characterizing RCFT. The polynomial equations involve the matrices S and T and two other matrices C and N expressing the duality properties of the tree-level conformal blocks. To exhibit the equations satisfied by C and N it is convenient to introduce chiral vertices (see, for example, Schroer 1987, and references therein; Fröhlich 1987; Tsuchiya & Kanie 1987). They are operators which represent the holomorphic three-point functions:

$$\Phi\{^{\ i}_{jk}\}(z): \quad H_k \to H_j, \tag{5}$$

where i is an index for a primary field. The conformal blocks can be written in terms of expectation values of products of chiral vertices. For example:

$$\mathscr{F}_p^{\ ijkl}(z, w) = \langle i|\, \Phi\{^{\ j}_{ip}\}(z)\, \Phi\{^{\ k}_{pl}\}(w)\, |l\rangle. \tag{6}$$

The matrix C describes the exchange of two chiral vertices. At the level of vertices it is the matrix representing the braiding (through analytic continuation) of the j, k legs of (6). Graphically,

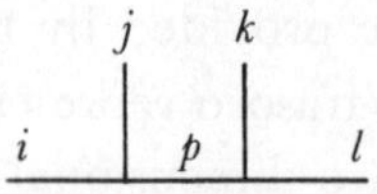

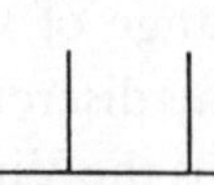

$$\tag{7}$$

where the left-hand side is a pictorial representation of the block, $\mathscr{F}_p^{\ ijkl}$. The matrix N is a consequence of the associativity of the operator product expansion:

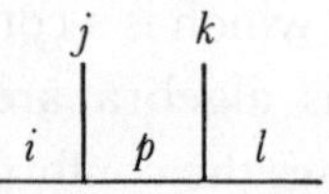

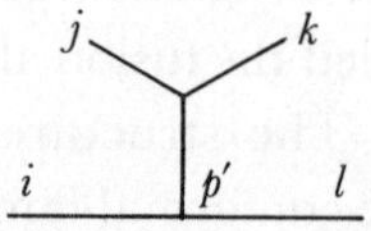

$$\tag{8}$$

The hexagon equation for C follows from the defining relations of the braid group, and the pentagon equation satisfied by N is a consequence of the associativity of the operator product expansion (see Moore & Seiberg (1989) for details). The proof of the Verlinde conjecture is a consequence of the pentagon identity after one writes a precise representation of the Verlinde operators. These are defined as follows. On the torus we choose a homology basis (a, b). For an RCFT given a primary field ϕ_i, we can always find a conjugate field $\phi_{\bar{\imath}}$ such that the operator product $\phi_i \times \phi_{\bar{\imath}}$ contains the identity. The Verlinde operators $\phi_i(a), \phi_j(b)$ act on the characters $\chi_i(\tau)$. They correspond to inserting the identity factorized into $\phi_i, \phi_{\bar{\imath}}$, then taking ϕ_i around the a- or b-cycle, and finally recombining the two operators into the identity once again. If the a-cycle represents the equal-time surface, the action of $\phi_i(a)$ on χ_j is diagonal:

$$\phi_i(a)\, \chi_j(\tau) = \lambda_i^{(j)}\chi_j(\tau). \tag{9a}$$

The action of $\phi_j(b)$ is more complicated;

$$\phi_i(a)\, \chi_j(\tau) = A_{ij}^{\ k}\chi_k(\tau), \tag{9b}$$

because a and b are exchanged by the modular transformation S, then $\phi_i(b) = S\phi_i(a)\, S^{-1}$, and the matrices $(A_i)_j^{\ k} \equiv A_{ij}^{\ k}$ all commute. In fact, S diagonalizes the A_is. The conjecture by Verlinde was that A_i and N_i coincide.

There are two ways that quantum groups (Drinf'eld 1986) enter into conformal field theory. First, the Verlinde operators are associated to closed paths on the torus. It is possible to construct analogues to these operators for open paths at tree level (Alvarez-Gaumé *et al.* 1989 *a*). The advantage is that these operators are compatible with the operation of taking traces, and they also imply the hexagon and pentagon equations, hence the proof of the Verlinde conjecture becomes conceptually simpler. Furthermore, these open Verlinde operators satisfy the defining relations for a quantum group. The second application of quantum groups has to do with the fact that they provide solutions to the polynomial equations (Alvarez-Gaumé *et al.* 1988). Moreover, all the known RCFTs can be obtained by using the Goddard–Kent–Olive (GKO) construction (Goddard *et al.* 1986) for appropriate groups G, H, $H \subset G$, and their C, N matrices and modular properties seem to follow from the quantum deformations of G and H (Alvarez-Gaumé *et al.* 1989 *b*). Whether the complete set of solutions to the polynomial equations are given by quantum groups is not clear at present.

In the first application we begin with the space of chiral vertices V compatible with the fusion rules $N_{ij}{}^k$. On V we can define two operations: one is sewing, denoted by *, and the other one is taking characters. We want to find an algebra of automorphisms of V compatible with (i) sewing, (ii) braiding, (iii) the N operation (8) or s–t duality. Denoting collectively by s the three labels in the chiral vertex (5), the braiding (or exchange) operation can be written as

$$\Phi_{S_1}(z)\,\Phi_{S_2}(w) = R_{S_1 S_2 S_3 S_4}\,\Phi_{S_3}(w)\,\Phi_{S_4}(z). \tag{10}$$

If Q is the automorphism algebra and $X \in Q$, we can write the action on V as

$$X(\Phi_{S_1}) = \sum_{S_2} C^{(1)}_{S_1 S_2}\,\Phi_{S_2}. \tag{11}$$

If we think of R as acting on $V \otimes V$, condition (i) becomes

$$RC_1 C_2 = C_2 C_1 R, \quad C_1 = C \otimes 1, \quad C_2 = 1 \otimes C, \tag{12}$$

reminiscent of the definition of a quantum group (Drinf'eld 1986). If all we had was (12), we could always define a representation of Q by using the adjoint representation (in analogy with ordinary Lie algebras). The problem is that the constraint imposed by the fusion rules,

$$\Phi\{{}^{ij}_k\}\,\Phi\{{}^l_{mn}\},$$

vanishes unless $k = l$. Nevertheless, for every primary field we can construct an element of Q. Choosing for simplicity a self-conjugate primary field (i.e. $\phi_a \times \phi_a = 1 + \ldots$) we obtain

$$X^a(\Phi\{{}^j_{ik}\}) = \sum_{m,n} R\left({}^a_{mi}\right)\left({}^j_{ik}\right)\left({}^j_{mn}\right)\left({}^a_{nk}\right)\Phi\{{}^j_{mn}\}. \tag{13}$$

Graphically, we can represent the chiral vertex as in figure 1, with the representations H_i and H_k attached in the open boundaries and the primary field ϕ_j at the point z. The operator (13) can be interpreted as in figure 2; we factorize the identity into ϕ_a and $\psi_{\bar a}$ and then move the fields to the i, k ends along the path shown. For more complicated surfaces one can do the same operation for each open path. This is why we can interpret X^a as the open Verlinde operators. The conditions (i), (ii) and (iii) imply also the tree-level polynomial equations (see Alvarez-Gaumé *et al.* 1989 *a* for details). Finally, the action on the characters is based on the operation of taking traces. In particular, the characters are, schematically,

$$\chi_i = \mathrm{Tr}_{H_i}\, q^{L_0} \underset{i\ i}{\perp}{}^{id}, \tag{14}$$

[27]

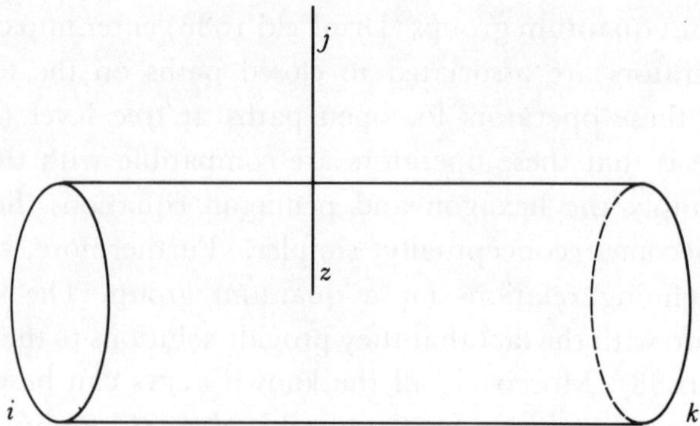

FIGURE 1. Graphical representation of the chiral vertex.

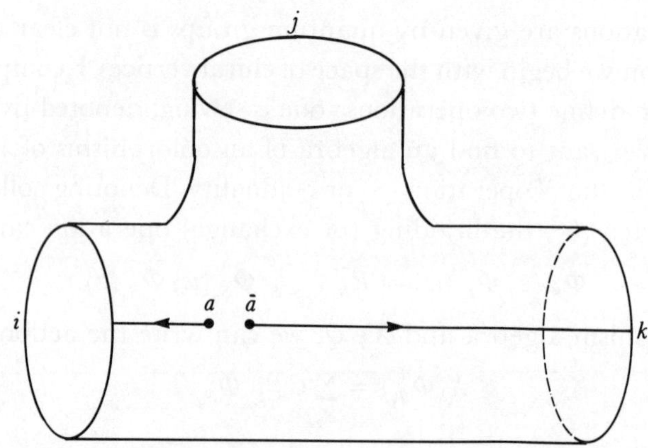

FIGURE 2. Interpretation of the operator defined by equation (13).

where we have represented the chiral vertex by $i \perp^j k$. The action of X^a on χ_i is defined by first acting on the chiral vertex and then taking the trace. It is easy to see, by using (13) and the triviality of braiding with the identity, that

$$X^a(\chi_j) = \sum_m N_{aj}{}^k \chi_k, \tag{15}$$

which is indeed the standard Verlinde operator.

The second use of quantum groups is related to the solutions of the polynomial equations. This can be illustrated with the quantum group $SL(2, q)$ which is associated to the level k Wess–Zumino–Witten theory (wzw) (see Alvarez-Gaumé et al. 1988). This algebra satisfies the defining relations:

$$[X^+, X^-] = (q^{\frac{1}{2}H} - q^{-\frac{1}{2}H})/(q^{\frac{1}{2}} - q^{-\frac{1}{2}}), \quad [H, X^\pm] = \pm 2X^\pm. \tag{16}$$

When q is an arbitrary real or complex number the representation theory of this algebra is analogous to the classical case. In RCFT, q is a root of unity. For instance, in the wzw theory, $q = \exp(2\pi i/k+2)$. Now the representation theory becomes more interesting. The only regular representations have spin $j = 0, \frac{1}{2}, 1, \ldots, \frac{1}{2}k$, as in the wzw theory. Furthermore, the composition of angular momentum generates the wzw fusion rules,

$$[j_1] \times [j_2] = \sum_{j=|j_1-j_2|}^{\min(j_1+j_2, K-j_1-j_2)} [j], \tag{17}$$

and the C and N matrices are given by the q-analogues of the $6-j$ symbols computed in Kirillov & Reshestikhin (1988),

$$C_{jj'}\begin{bmatrix} j_2 & j_3 \\ j_1 & j_4 \end{bmatrix} = (-1)^{j+j'-j_1-j_4} q^{\frac{1}{2}(C_{j_1}+C_{j_4}-C_j-C_{j'})} \begin{Bmatrix} j_2 & j_1 & j \\ j_3 & j_4 & j' \end{Bmatrix}_q,$$

$$N_{jj'}\begin{bmatrix} j_2 & j_3 \\ j_1 & j_4 \end{bmatrix} = \begin{Bmatrix} j_1 & j_2 & j \\ j_3 & j_4 & j' \end{Bmatrix}_q.$$

In this way, one reproduces the results of Schroer (1987), Fröhlich (1987) and Tsuchiya & Kanie (1987), and, by using the q-characters of $SL(2, q)$, one finds that the modular transformations are represented by $q \to q^{-1}$.

A plausible reason why the quantum group appears is because the polynomial equations are concerned with the Hilbert space of the theory modulo the chiral algebra. In the Kac–Moody case relevant for the wzw theory this means roughly that we forget the moding in the Kac–Moody generators. Naïvely, one might expect that after doing this one is left with the classical algebra. The deformation, however, is a consequence of the central extension of the Kac–Moody algebra which is crucial in determining its representations.

It is quite plausible that all solutions to the polynomial equations are given by combinations of quantum groups. Work in this direction is in progress.

I thank the organizers of the meeting for the opportunity to present this material in such a stimulating environment.

References

Alvarez-Gaumé, L., Gomez, C. & Sierra, G. 1988 The quantum group interpretation of some conformal field theories. CERN Preprint TH. 5267.

Alvarez-Gaumé, L., Gomez, C. & Sierra, G. 1989*a* Hidden quantum symmetries in rational conformal field theories. *Nucl. Phys.* (Submitted.)

Alvarez-Gaumé, L., Gomez, C. & Sierra, G. 1989*b* (In the press.)

Belavin, A. A., Polyakov, A. M. & Zamolodchikov, A. 1988 *Nucl. Phys.* B **241**, 333.

Drinf'eld, V. G. 1986 Quantum groups. In *Proc. Int. Cong. Mathematicians*, MSRI, Berkeley.

Friedan, D., Qiu, Z. & Shenker, S. 1984 *Phys. Rev. Lett.* **52**, 1575.

Fröhlich, J. 1987 Statistics of fields, the Yang–Baxter equations and the theory of knots and links. 1987 Cargèse Lectures.

Goddard, P., Kent, A. & Olive, D. 1986 *Communs math. Phys.* **103**, 105.

Kirillov, A. N. & Reshestikhin, N. Y. 1988 LOMI Preprint E-9-88.

Moore, G. & Seiberg, N. 1989 *Phys. Lett.* (In the press.)

Schroer, B. 1987 Algebraic aspects of non-perturbative quantum field theories. Como Lectures, August 1987.

Tsuchiya, A. & Kanie, Y. 1987 *Lett. math. Phys.* **13**, 303.

Verlinde, E. 1988 *Nucl. Phys.* B **300**, 360.

Witten, E. 1984 *Communs math. Phys.* **92**, 455.

Phil. Trans. R. Soc. Lond. A **329**, 349–357 (1989)
Printed in Great Britain

The search for higher symmetry in string theory

By E. Witten

*School of Natural Sciences, Institute for Advanced Study, Olden Lane, Princeton,
New Jersey 08540, U.S.A.*

Some remarks are made about the nature and role of the search for higher symmetry
in string theory. These symmetries are most likely to be uncovered in a mysterious
'unbroken phase', for which $(2+1)$-dimensional gravity provides an interesting and
soluble model. New insights about conformal field theory, in which one gets 'out of
flatland' to see a wider symmetry from a higher-dimensional vantage point, may offer
clues to the unbroken phase of string theory.

1. Search for the unbroken phase

I thought I might begin by stating my general outlook on string theory.

In my opinion, the basic challenge in string theory is not, as sometimes said, to 'understand
non-perturbative processes in string theory'. In fact, the basic problem does not lie in the
quantum domain at all. The basic problem, now and probably for many years to come, is to
understand the classical theory properly.

Understanding the classical theory means above all understanding the geometrical ideas
that parallel those of general relativity. It is indeed a key aspect of the fascination of string
theory that in string theory, general relativity is modified *even in the classical domain*. The key
geometrical ideas upon which string theory should be founded must be worthy successors to the
principle of equivalence and riemannian geometry, and they must lead to what we now know
of as string theory as ineluctably as the principle of equivalence leads to general relativity.

Not only is the search for geometric understanding crucial, it is the key area where progress
can be expected in the next stage of development. The advance in viewpoint that we need is
probably big, but perhaps no bigger than the advance made long ago when 'dual models' were
reinterpreted as 'string theories'. I suppose my basic outlook on physics is that conceptual
problems such as those cited above have answers, and these answers are eventually found by
human beings. One might not say the same for dynamical problems.

So the central task is to

  (i) find the right degrees of freedom;

  (ii) find the right invariance group; and

  (iii) formulate the right lagrangian.

These are, of course, the usual questions of string field theory, and I consider them to be the
right questions though string field theory as we know it is very likely not based on the right
degrees of freedom.

In thinking about these matters, it is sobering to realize how little we know with certainty.
Thus it is often said that 'string theory is a theory of extended objects', and Riemann surfaces
are often felt to play a crucial role. But the $1/N$ expansion of quantum chromodynamics (QCD)
may offer a cautionary tale. As we have long known, QCD with $SU(N)$ gauge group has an

[31]

expansion formally very similar to the topological expansion of string theory, with mesons and glueballs playing the role of open and closed strings, and a mesh of planar gluons playing the role of the string world sheet. In this picture, it is confinement of colour that creates effective 'Riemann surfaces'. We do not know of a simple Veneziano-like description of the $1/N$ expansion of QCD, but one may exist. (And if there is an area of QCD theory where a real breakthrough can be made, this is it. Certainly, the search for such a description has been one of the main motivations of the Soviet school of string theorists, throughout the 1980s.)

QCD from this viewpoint looks like a 'theory of extended objects', all based on Riemann surfaces. Yet conceptual understanding from this point of view is probably hopeless. It is necessary to learn that the correct degrees of freedom are quarks and gluons, before learning that the correct notion is gauge theory.

At least in thought experiments, one can neatly see the world of extended objects and Riemann surfaces come to an end in QCD, by heating to the deconfinement temperature, T_{dec}. Since the early work on the deconfinement transition (Polyakov 1978; Susskind 1979) it has been suspected that a similar phenomenon may be occurring in string theory at the so-called Hagedorn temperature, T_{H}. There are a variety of reasons to suspect that this may be so. One argument that I find fairly convincing is that in string theory, the free energy begins in genus one if $T \leqslant T_{\text{H}}$, but receives a genus zero contribution if $T \geqslant T_{\text{H}}$ (Atick & Witten 1988). This parallels an analogous behaviour in QCD.

Another reason to suspect the occurrence in string theory of an analogue of the deconfinement transition is that, on the face of it, the building blocks that we are currently using in our attempts to describe string theory do not seem adequate for the job. String fields are too messy, and two-dimensional field theory is too singular. Let me pause for a word on the latter point. We recall that unitary conformal field theories come in finite-parameter families (the number of parameters being the number of relevant operators). This is a very different sort of thing from what we want to describe time-dependent processes, where it should be possible to excite arbitrary massive modes of the string. I believe that this is a crucial issue, and in no way a technicality. In any serious attempt to study lorentzian signature sigma models with time-dependent fields, the singularities of quantum field theory will appear with a vengeance, with operators of arbitrarily negative dimension appearing on the scene. It is significant that this subject has been so little investigated in print.

General relativity should again offer the paradigm of what we want. The basic concept of a metric is flexible and elementary. Choosing a metric does not require solving any equations. We should expect to one day develop string theory with tools as flexible and elementary as the concept of a riemannian metric. This may be possible only by understanding the 'unbroken phase' of string theory, with the higher symmetries restored. What properties do we expect of this phase?

There are many reasons to believe that in string theory there is no such thing as distances less than the fundamental length $\sqrt{\alpha'}$. Among the reasons for believing this are the duality in distance between radius R and radius α'/R, the fact that (as in the paper by David Gross, this Symposium) one cannot probe distances below $\sqrt{\alpha'}$ by using high-energy, fixed-angle scattering, and the fact that the growth of the free energy above the Hagedorn temperature seems to be slower than in field theory. What one might imagine would be a world in which at distances above $\sqrt{\alpha'}$, normality prevails, but at distances below $\sqrt{\alpha'}$, not just physics as we know it but local physics altogether has disappeared. There will be no distance, no times, no energies, no particles, no local signals – only differential topology, or its string theoretic successor.

To clarify this a bit, let us note that one often thinks of four-dimensional general relativity, in the phase we live in, as a system in which the space-time symmetries are unbroken. On reflection, though, one sees that the non-zero expectation value of the metric, $g_{\mu\nu} \sim \eta_{\mu\nu}$, violates *local* general covariance and leaves only the *global* Poincaré symmetries. To restore local general covariance, one would need $\langle g_{\mu\nu} \rangle = 0$. But in the absence of an expectation value of the metric, there is no notion of distance, time or energy, and there is no way to have particles or local signals because there is no way for these to stay inside the light cone.

In general relativity, we are usually satisfied with unbroken global symmetry. We do not usually hunger to restore the local symmetry and thus bring physics to an end. In string theory, however, the graviton is just one of a tower of states. Our crucial task is to understand the symmetries associated with the massive states, the masses presumably arising by a Higgs-like mechanism. In string theory, an expectation value of the metric breaks the higher symmetries even as global symmetries. To restore the higher symmetries of the string even as global symmetries, one needs $\langle g_{\mu\nu} \rangle = 0$, and in this case general covariance, and the whole higher invariance group of which it is part, is restored as a *local* symmetry. As we have already noted, restoration of general covariance makes local physics impossible, and this is why I believe that string theory must be understood in a framework in which locally there is nothing there.

The most concrete picture I can offer of such local symmetry restoration comes in $(2+1)$-dimensional gravity. Before entering into this, let us ask why $(3+1)$-dimensional gravity is unrenormalizable. I write the lagrangian in vierbein formalism:

$$\mathscr{L} = \int e^a \wedge e^b \wedge (\mathrm{d}\omega + \omega \wedge \omega)^{cd} \cdot \epsilon_{abcd}. \tag{1.1}$$

(Here e^a is the vierbein, ω^{ab} the spin connection, and ϵ_{abcd} the alternating symbol with $\epsilon_{0123} = +1$.) Note that if we regard e and ω as fields of dimension one (this is conventional for ω, but e is usually regarded as a field of dimension zero), then every term in (1.1) has dimension four, so we might think that general relativity is renormalizable in four dimensions. This is not so; where is the mistake?

The error is that if e and ω have positive dimension, the short distance behaviour must be governed by $e = \omega = 0$. This is what I mean, in general relativity, by the 'unbroken phase' with *local* general covariance restored. But we do not understand four-dimensional general relativity in the 'unbroken phase'. Mathematically, we cannot expand (1.1) around $e = \omega = 0$ because there is no quadratic term. Physically, gravitons could not propagate in such a region, so in four dimensions the interface between the broken and unbroken phases must be quite subtle.

In $2+1$ dimensions, the story is different. Once initial reluctance is overcome, the action,

$$\mathscr{L} = \int (e^a \wedge (\mathrm{d}\omega + \omega \wedge \omega)^{bc} + \lambda e^a \wedge e^b \wedge e^c) \, \epsilon_{abc} \tag{1.2}$$

(I have incorporated a cosmological constant term), can readily be expanded around $e = \omega = 0$. What is more, the resulting expansion (if placed in the framework of gauge theory, as is made possible by observations in (Achúcarro & Townsend 1986; Witten 1989a) is renormalizable by power counting, and is easily seen to be finite. Given that (1.2) makes sense and is finite in the unbroken phase, it can be expanded around any classical solution without spoiling the finiteness.

[33]

24-2

Macroscopically, e and ω can be given the usual riemannian interpretation of general relativity. Microscopically, this is not so. After all, the short-distance régime is an expansion around $e = 0$, and a space-time point with $e = 0$ (or even $\det e = 0$) is a singularity from the standpoint of riemannian geometry. In a gauge theory interpretation, we do not regard such occurrences as singularities, and this is a crucial departure from the riemannian interpretation. Actually, in the gauge theory interpretation, e and ω can be gauged away locally, so in a very precise sense there is 'nothing' locally. Nevertheless, in the large, one has macroscopic space-times. (These macroscopic space-times are empty, and have only global dynamics, because there are no gravitons in $2+1$ dimensions. It is this that makes $(2+1)$-dimensional gravity simple, while the incorporation of the unbroken phase will be vastly more difficult in $3+1$ dimensions.)

I would like to believe that something like this occurs in nature, but far more subtle, because there are signals propagating in the 'broken phase' and the interface between them must be at least as subtle as in QCD.

In seeking such an understanding, we are in the situation of the proverbial man who has lost his key and is looking for it in the dark. There is little to do except to look under a well-lit lamp-post and hope for the best. But it is up to us to pick a promising lamp-post to look under. The missing key is the geometrical key to string theory. My own favourite lamp-post – about which I will say a few words in the next section – is the investigation of geometrically interesting structures associated with quantum field theory, in dimensions ranging from two (the world-sheet) to four (the dimension of space-time, at least macroscopically). These are the most interesting dimensions in geometry, topology and physics. The idea that to understand two-dimensional field theory one must get out of 'flatland' and find a higher-dimensional perspective was originally advocated by Sir Michael Atiyah; I am grateful for the inspiration he has provided.

2. Higher symmetries in conformal field theory

Let us first recall that in *four* space-time dimensions, the *spin* of a massless particle is a representation of $SO(2)$ and so is an integer or half-integer. Thus fermions, gauge bosons and gravitons may have *positive* or *negative* helicity. The chiral decomposition for fermions is one of the central facts in particle physics, but this is much less true for bosons, though the polarization of light is certainly one of the basic observations in electromagnetism, and in QCD the self-dual Yang–Mills equations,

$$F_{\mu\nu} = \tilde{F}_{\mu\nu}, \tag{2.1}$$

do enter. The chiral decomposition is less important for bosons because it cannot be made locally.

Twistor theory (for introductory articles see Lerner & Sommers (1978)), in which four-dimensional space-time R^4 is replaced by a higher-dimensional space CP^3, exhibits a remarkable higher symmetry of the chiral or self-dual theory. All information is encoded globally. The challenge in twistor theory has always been to combine the left-handed and right-handed theories. (A proposal was made in Witten (1978) and Isenberg *et al.* (1978). It remains to be seen if this proposal can offer any guidance for conformal field theory.) This is a difficult problem, because the Einstein and Yang–Mills equations are not 'rational' in the language of conformal field theory.

The chiral decomposition in four dimensions has, of course, an analogue in two dimensions in the decomposition in left-movers and right-movers. This decomposition is powerful primarily in the so-called rational conformal field theories. Its use in that context is practically the only truly two-dimensional tool we know of in conformal field theory. When this tool is not available, we know scarcely more about conformal field theory in two dimensions than in four dimensions.

One way that the chiral decomposition enters in conformal field theory is this. Let us work on a Riemann surface Σ of genus g, with the moduli denoted as m^i. A conformal field theory is said to be 'rational' if its partition function $Z(m^i, \bar{m}^j)$ has an expansion,

$$Z(m^i, \bar{m}^j) = \sum_{\lambda=1}^{k} f_\lambda(m^i) \tilde{f}_\lambda(\bar{m}^j), \qquad (2.2)$$

as a sum of products of holomorphic functions f_λ and anti-holomorphic ones $\tilde{f}_\lambda$. If this is so, the $\{f_\lambda(m^i)\}$ span a finite-dimensional vector space $\mathscr{H}_\Sigma$ canonically associated with Σ. Σ may now be regarded purely as an oriented, topological surface; the moduli have been buried as arguments of the $f_\lambda(m^i)$, and forgetting the origin of the f_λs, we now merely think of them as a basis of a vector space $\mathscr{H}_\Sigma$.

Friedan & Shenker (1987) urged us to consider the behaviour of the $\mathscr{H}_\Sigma$ as Σ degenerates. For instance, we consider the degeneration of a surface Σ of genus g to surfaces Σ_1, Σ_2 of genus g_1 and g_2 (with $g_1 + g_2 = g$). As indicated in figure 1, it is natural to think of such a process as a time-dependent process in which a surface of genus $g_1 + g_2$ enters in the past and two surfaces of genus g_1 and g_2 emerge in the future. Such a space-time history naturally gives a three-manifold. To begin with, it appears that on this three-manifold there is a preferred 'time' coordinate, as it really represents a one-parameter family of two-dimensional configurations. I claim, however, that this is an illusion, and that one can consider the picture sketched in figure 1 to have full three-dimensional symmetry.

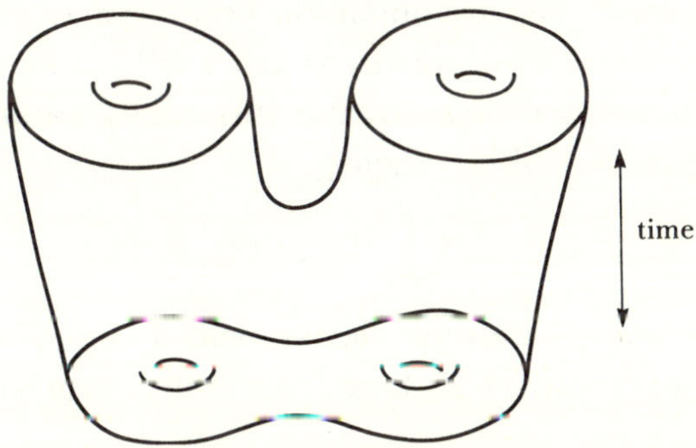

FIGURE 1. A Riemann surface of genus two breaking into two surfaces of genus one.

To further elaborate this claim, recall that in such theories the *primary fields* $\Phi_i(z, \bar{z})$ have decompositions

$$\Phi_i(z, \bar{z}) = \varphi_i(z) \tilde{\varphi}_i(\bar{z}), \qquad (2.3)$$

where the $\varphi_i(z)$ are holomorphic but not quite local. In fact, the n-point correlation function,

$$G_{i_1, \ldots, i_n}(z_{i_1}, \ldots, z_{i_n}) = \langle \varphi_{i_1}(z_1) \ldots \varphi_{i_n}(z_n) \rangle, \qquad (2.4)$$

is a multivalued function with, say, k branches, which we will call $G^\sigma_{i_1, \ldots, i_n}, \sigma = 1, \ldots, k$.

Forgetting about the origin of the G^σs and just thinking of them as a basis of a vector space, we get a vector space $\mathcal{H}_{\Sigma, i_1, \ldots, i_n}$ that depends on the surface Σ as a topological surface with n points labelled as $i_1, \ldots, i_n$. The moduli z^k are irrelevant; they have disappeared as arguments of the G^σ.

When studying a multivalued function, such as the correlation function (see equation (2.4)), it is important to study the monodromies, that is, the linear transformations of the branches G^σ that occur when the points z_i loop around one another. It is convenient to think of the z_i as n points on the Riemann sphere (or complex plane C), that loop around one another in the course of time, forming a braid (figure $2a$) in three-space. In figure $2a$, only the topology of the braid is relevant. Thus one may make arbitrary time-dependent diffeomorphisms of the plane C in which the points are moving in figure $2a$. I claim, though, that there is a higher, three-dimensional symmetry in figure $2a$. In fact, if one glues together the top and bottom of the braid under discussion, one forms (see figure $2b$) a knot. It turns out that one should think of this knot as an intrinsically three-dimensional object; that is, we now allow full three-dimensional diffeomorphisms.

(a)   (b)

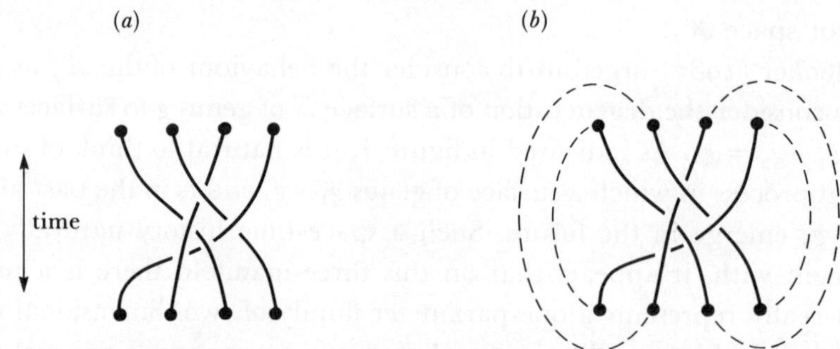

FIGURE 2. A braid (a) whose ends can be joined to make a knot (b).

To justify these assertions, one must obtain the vector spaces $\mathcal{H}_\Sigma$ and $\mathcal{H}_{\Sigma, i_1, \ldots, i_n}$ from a *three-dimensional viewpoint*. We have been urged (Atiyah 1988) to look to quantum field theory to provide this framework, and for physicists this is certainly natural.

Of course, a three-dimensional lagrangian,

$$\mathcal{L} = \int_M \mathrm{d}x\,\mathrm{d}y\,\mathrm{d}t\, W(\phi_i, \partial\phi_j) \tag{2.5}$$

(M denotes space-time and ϕ_i are some fields), will, upon quantization, associate a Hilbert space $\mathcal{H}_\Sigma$ to every Riemann surface Σ. If $\mathcal{L}$ is generally covariant, $\mathcal{H}_\Sigma$ will depend on Σ only as a *topological* surface.

So to understand the three-dimensional symmetry of two-dimensional conformal field theory, we look at generally covariant three-dimensional quantum field theories. General covariance does not allow us to introduce an *a priori* metric on M. If there is a metric at all, it must be a dynamical variable. But if so, the quantum wave-functions will depend upon the metric, and the quantum Hilbert spaces will be infinite dimensional. As we want finite-dimensional spaces (to agree with rational conformal field theory), we must avoid this, so we must consider three-dimensional theories in which general covariance is achieved with *no metric at all*. Thus we must relate two-dimensional conformal field theory to the *unbroken phase* of three-dimensional gravity, in which the metric is irrelevant.

[36]

The basic example seems to be Chern–Simons gauge theory. So (Witten 1989 b) we pick a compact gauge group G and a positive integer k. We introduce a gauge field $A_i^a(x)$ and write the lagrangian

$$\mathscr{L} = \frac{k}{4\pi} \int_M \mathrm{Tr}\,(A \wedge \mathrm{d}A + \tfrac{2}{3}A \wedge A \wedge A). \tag{2.6}$$

No metric is needed, so this is generally covariant.

The reason that the quantum field theory with this lagrangian is exactly soluble is that this theory is trivial locally. For instance, at the classical level the Euler–Lagrange equation for this theory is $F_{ij}^a = 0$, and implies that A_i can be gauged away locally. Thus as in twistor theory, all information is to be encoded globally.

Quantum mechanically, to construct the Hilbert space $\mathscr{H}_\Sigma$ associated with a riemannian surface Σ, we pick the gauge $A_0 = 0$. The Gauss law constraint $0 = \delta\mathscr{L}/\delta A_0^a = F_{12}^a$ then tells us to consider only flat corrections. After dividing by gauge transformations, the phase space that must be quantized is what I shall call $\mathscr{M}_\Sigma$, the moduli space of flat G corrections on Σ, up to gauge equivalence. $\mathscr{M}_\Sigma$ is a compact symplectic manifold, and its quantization gives the mysterious Hilbert spaces $\mathscr{H}_\Sigma$ of G *current algebra at level k*. If one repeats this in the presence of static charges in representations $R_{i_1}, \ldots, R_{i_n}$ of G, one gets more general Hilbert spaces $\mathscr{H}_{\Sigma, i_1, \ldots, i_n}$ of multivalued correlation functions (see equation (2.4)) with Φ_i being a current algebra primary field in the representation R_i. In turn, the latter spaces are essentially the Jones braid representations (Jones 1985) that are associated with the celebrated Jones polynomial of knot theory; the relation of these spaces to conformal field theory was first perceived by Tsuchiya & Kanie (1988).

One might view these mysterious Hilbert spaces $\mathscr{H}_\Sigma$ as rather esoteric objects, but in fact, the whole current algebra theory in two dimensions can be reconstructed from the topological theory in three dimensions. As explained in the last few pages of (Witten 1989), this is done by quantizing the Chern–Simons theory on *Riemann surfaces with boundary*. For instance, we take Σ to be a disk D. Taking M to be $D \times R^1$, with R^1 denoting 'time', the lagrangian (see equation (2.6)) is only gauge invariant under the group G_1 consisting of gauge transformations that are 1 on the boundary of D. Quantizing in $A_0 = 0$ gauge, the Gauss law constraint still implies $F_{12}^a = 0$, so (as $D \times R^1$ is simply connected) one has

$$A_i(x, y) = \partial_i U \cdot U^{-1}, \tag{2.7}$$

where

$$U(x, y) \approx U(x, y) \cdot V \tag{2.8}$$

for any constant group element V (because $U \to UV$ leaves A unchanged in (equation (2.7)). Up to gauge transformation by G_1, only the value of U on the boundary of D is relevant. The boundary of D is a circle S^1, which we may parametrize by an angle θ, $0 \leqslant \theta \leqslant 2\pi$. So $U(\theta)$ defines a point in the loop space $\mathscr{L}G$ of G; more exactly, in view of the equivalence of equation (2.8), we should think of $U(\theta)$ as a point in the homogeneous space $\mathscr{L}G/G$. Thus, it is $\mathscr{L}G/G$ that is to be quantized. According to Segal (1981), quantization of $\mathscr{L}G/G$ gives the irreducible (vacuum) representation of G current algebra at level k.

More generally, by quantizing on an n-holed surface of genus g (figure 3) one can study the n-point functions in genus g from the three-dimensional viewpoint.

It is not difficult to show abstractly that any rational conformal field theory gives rise automatically to a three-dimensional generally covariant theory. One approach to proving this

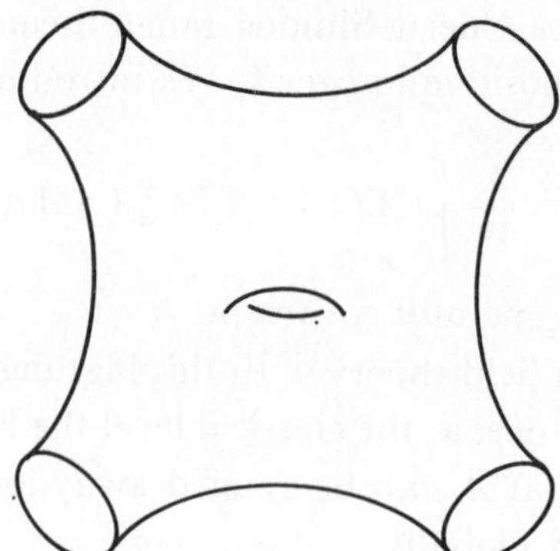

FIGURE 3. A surface of genus one with four holes.

(suggested to me by W. Thurston) involves decomposing a three manifold via a Morse function, by using the behaviour of the $\mathcal{H}_\Sigma$ under degeneration of Σ to define transition amplitudes in pictures such as that of figure 2, and then verifying that the choice of Morse function does not matter. Details will be described elsewhere.

More concretely, though, Moore & Seiberg (1989) have recently shown that G/H models may also be interpreted as Chern–Simons theories of an appropriate group (essentially G × H). It is easy to see that this is also true of rational orbifolds, and of certain generalizations of G/H models that will be described elsewhere. So it is possible to conjecture that arbitrary rational conformal field theories in two dimensions can be derived from Chern–Simons theories in three dimensions.

Actually one can make a stronger conjecture along these lines. Apart from the models that are related to two-dimensional conformal field theories, the only three-dimensional generally covariant theories I know of that at first sight appear not to be Chern–Simons theories are general relativity and the three-dimensional reduction of Donaldson theory. But both three-dimensional gravity (Achúcarro & Townsend 1986; Witten 1989a) and the three-dimensional reduction of Donaldson theory (Witten 1989c) have turned out to be Chern–Simons theories. So one may conjecture that every three-dimensional generally covariant theory is actually a Chern–Simons theory of some group or supergroup, not necessarily connected or simply connected. Perhaps this conjecture should be stated only in the context of the unbroken phase of general relativity.

Integrable lattice models in 1+1 dimensions are also intimately connected with three-dimensional quantum field theory, as is apparent from the literature of recent years on knot theory and statistical mechanics, along with the above observations. I will try elsewhere (Witten 1989d) to elucidate at least part of this connection from a physical point of view.

In conclusion, if one considers rational conformal field theories and their cousins, integrable systems, in two dimensions, one sees that these are extremely rich systems. There are an incredible variety of extremely rich facts, related to each other in an incredible diversity of ways. To bring order to this chaos looks hopeless. But by stepping out of flatland and looking at things from the vantage point of three dimensions, one can find a more powerful viewpoint, where rational and integrable systems can be derived from a subtler and more incisive starting point. This step out of flatland, to a higher vantage point from which wider symmetry can be seen, is temptingly akin to what we need in string theory. What is more, we have been urged

(Atiyah 1988) to take yet another step to four dimensions, the most physical dimension, the richest dimension for geometry, and the critical dimension for quantum field theory.

This research was supported in part by NSF grant 86-20266 and NSF Waterman Grant 88-17521.

REFERENCES

Achúcarro, A. & Townsend, P. 1986 *Phys. Lett.* B **180**, 89.

Atick, J. & Witten, E. 1988 The Hagedorn transition and the number of degrees of freedom in string theory. *Nucl. Phys.* B **310**, 291.

Atiyah, M. F. 1988 New invariants of three and four dimensional manifolds. In *The mathematical heritage of Hermann Weyl, Proc. Symp. pure Math.* **48** (ed. R. O. Wells, Jr). Providence: AMS.

Friedan, D. & Shenker, S. 1987 *Nucl. Phys.* B **281**, 509.

Isenberg, J., Yasskin, P. & Green, P. 1978 *Phys. Lett.* B **78**, 462.

Jones, V. F. R. 1985 A polynomial invariant for links via subfactors. *Bull. AMS* **12**, 103.

Lerner, M. & Sommers, D. (ed.) 1978 *Complex manifold techniques in theoretical physics.* Pitman.

Moore, G. & Seiberg, N. 1989 Taming the conformal zoo. IAS preprint HEP-89/6.

Polyakov, A. M. 1978 *Phys. Lett.* B **72**, 477.

Segal, G. 1981 Unitary representations of some infinite dimensional groups. *Communs math. Phys.* **80**, 301.

Susskind, L. 1979 *Phys. Rev.* D **20**, 2610.

Tsuchiya, A. & Kanie, Y. 1988 In *Conformal field theory and solvable lattice models. Adv. Studies pure Math.* **16**, 297.

Witten, E. 1978 An interpretation of classical Yang–Mills theory. *Phys. Lett.* B **77**, 394.

Witten, E. 1989*a* 2+1 Dimensional gravity as an exactly soluble system. *Nucl. Phys.* B **311**, 46.

Witten, E. 1989*b* Quantum field theory and the Jones polynomial. *Communs math. Phys.* **121**, 351.

Witten, E. 1989*c* Topology changing amplitudes in 2+1 dimensional gravity. *Nucl. Phys.* B. (In the press.)

Witten, E. 1989*d* Gauge theories and integrable lattice models. *Nucl. Phys.* B. (In the press.)

Phil. Trans. R. Soc. Lond. A **329**, 359–371 (1989)

Printed in Great Britain

The search for a realistic superstring vacuum

By J. H. Schwarz

California Institute of Technology, Pasadena, California 91125, *U.S.A.*

Recent progress in understanding heterotic string compactification utilizes abstract algebraic methods. In particular, Gepner has given a prescription for constructing exactly soluble Calabi–Yau compactifications by using $N = 2$ minimal models. A large class of new $N = 2$ models, discovered by Kazama & Suzuki, can also be used. Recent progress in understanding $N = 2$ models and Calabi–Yau spaces by using mathematical techniques of singularity theory improves the prospects of a complete classification. A key question is whether a classical solution can give a reasonable first approximation to the exact quantum ground state even though string theory is strongly coupled and the perturbation expansion diverges.

Introduction

In the four years that have passed since string theory emerged as a popular approach to unification we have learned a great deal about many facets of the programme (Green *et al.* 1987). Two general categories of issues have emerged as critical and received much scrutiny in recent times. One is the quest for a genuine non-perturbative formulation of quantum string theory. A very broad range of ideas and approaches have been put forward, but as yet none has emerged as the consensus favourite. The second category of questions, which is the subject of this review, concerns the search for classical solutions of the heterotic string theory, with particular emphasis on those that are most promising for achieving phenomenological success.

Four years ago string theory was touted for its uniqueness. In a certain sense that is still a viable point of view today. There are only three theories – type I, type II, and heterotic – that appear to be internally consistent when analysed in perturbation theory. Each is completely free of parameters or other arbitrariness. What has become clear in the intervening years is that this uniqueness at the level of the fundamental equations (whatever they are) is not reflected at the level of classical solutions. A bewildering proliferation of classical solutions have been discovered by a variety of techniques. However, there is a unifying principle. Four dimensions can be taken to be flat Minkowski space with the remaining degrees of freedom described by an arbitrary conformal field theory having suitable central charges, supersymmetry and modular invariance. All classical solutions that have been proposed, whether or not expressed in these terms, can be interpreted as particular constructions of the internal conformal field theory.

Unfortunately, at the level of perturbative analysis, this programme has some disturbing arbitrariness. There is no compelling theoretical reason to separate off four-dimensional space-time or to require that it be a Minkowski space. Moreover, the possibilities for the internal conformal field theories are very numerous. The number can be reduced by imposing some physical requirements such as $N = 1$ supersymmetry (to solve the hierarchy problem) or a particular number of families.

One of the first proposals, Calabi–Yau compactification (Candelas *et al.* 1985), still looks the most promising. From the abstract conformal field theory point of view it corresponds to a class of $(2,2)$ superconformal models with central charge $c = 9$. The abstract approach treats related orbifold compactifications as part of the same category. The detailed procedure for turning an arbitrary $(2,2)$ superconformal model with $c = 9$ into an $N = 1, D = 4$ heterotic string solution with E_6 families of quarks and leptons has been worked out by Gepner ($1987b$).

Most of this review is concerned with the description of various techniques that have been developed for constructing $(2,2)$ superconformal models. Many examples, including the minimal models considered by Gepner, can be constructed from Kac–Moody algebra cosets, by using the method of Goddard *et al.* (1985). A supersymmetric extension of this method has been formulated by Kazama & Suzuki (1988) and applied to the construction of many new $N = 2$ superconformal models, which can also be used in the construction of the internal $c = 9$ model.

Gepner gave convincing evidence that certain of his models correspond to known Calabi–Yau compactifications. This seemed miraculous at the time, because his construction is entirely algebraic and uses exactly soluble minimal models. Calabi–Yau spaces on the other hand are very complicated geometric structures, none of which has a known metric. The origins of this 'miracle' are much better understood now in view of recent developments applying the mathematical techniques of singularity theory to the description of $N = 2$ models.

CONFORMAL FIELD THEORY

This section briefly reviews some of the central ideas in conformal field theory that are required in the sequel. For more thorough and systematic discussions the reader is referred to the review articles (Peskin 1987; Banks 1987; Ginsparg 1988).

It is extremely convenient to make a Wick rotation $\tau \to i\tau$ so as to euclideanize the string world-sheet and thereby make the metric $h_{\alpha\beta}$ positive definite. Having done this, we can introduce complex coordinates (in local patches)

$$z = e^{\tau + i\sigma} \quad \text{and} \quad \bar{z} = e^{\tau - i\sigma} \tag{1}$$

and regard the world sheet as a Riemann surface. The gauge invariances $\tau \pm \sigma \to f_\pm(\tau \pm \sigma)$ become conformal mappings $z \to \tilde{z}(z)$ and similarly for $\bar{z}$. Thus we are led to consider conformally invariant two-dimensional field theory (Belavin *et al.* 1984). These transformations are generated by the energy–momentum tensor components $T(z)$ and $\bar{T}(\bar{z})$.

For a conformal dimension h field we write

$$\phi(z) = \sum_{-\infty}^{\infty} \frac{\phi_n}{z^{n+h}}. \tag{2}$$

More generally, conformal fields are functions of z and $\bar{z}$ and thus have a pair of dimensions $(h, \bar{h})$. However, to simplify writing I will usually suppress $\bar{z}$ dependences. If the ghosts are included (so that the conformal anomaly cancels), $T(z)$ has dimension $(2,0)$ and $\bar{T}(\bar{z})$ has dimension $(0,2)$.

Under a finite transformation $z \to \tilde{z}(z)$, a conformal dimension h field transforms as follows

$$\phi(z) \to (\partial \tilde{z}/\partial z)^h \tilde{\phi}(\tilde{z}). \tag{3}$$

[42]

One sometimes says that ϕ is an 'h-form', because $\phi(z)\,(\mathrm{d}z)^h$ is invariant. The infinitesimal transformation is determined by the operator product expansion (OPE)

$$T(z)\,\phi(w) \sim h\phi(w)/(z-w)^2 + \partial\phi(w)/(z-w) + \dots. \tag{4}$$

The symbol ∂ means differentiation with respect to w, of course. The dots represent non-singular terms. The $N=1$ superconformal algebra corresponds to the OPEs

$$\left.\begin{aligned}
T(z)\,T(w) &\sim \frac{c}{2(z-w)^4} + \frac{2T(w)}{(z-w)^2} + \frac{\partial T(w)}{z-w} + \dots, \\[2mm]
T(z)\,G(w) &\sim \frac{3}{2}\frac{G(w)}{(z-w)^2} + \frac{\partial G(w)}{z-w} + \dots, \\[2mm]
G(z)\,G(w) &\sim \frac{2c}{3(z-w)^3} + \frac{2T(w)}{z-w} + \dots.
\end{aligned}\right\} \tag{5}$$

Thus $G(z)$ has dimension $(\tfrac{3}{2},0)$.

Particularly interesting examples of conformal fields are the two-dimensional currents associated with a Lie group symmetry in a conformal field theory (Goddard & Olive 1986). Using current conservation one can show that there is a holomorphic component $J^A(z)$ and an antiholomorphic component $\bar{J}^A(\bar{z})$, just as for T and G. As above, we consider $J^A(z)$ only. The zero modes J_0^A are the generators of a Lie algebra G with

$$[J_0^A, J_0^B] = \mathrm{i}f_{ABC}\,J_0^C. \tag{6}$$

The algebra of the currents $J^A(z)$ is an infinite-dimensional extension of this, known as an affine Lie algebra or as a Kac–Moody algebra $\hat{G}$. These currents have conformal dimension $h=1$. The Kac–Moody algebra is given by the OPEs

$$J^A(z)\,J^B(w) \sim \frac{k\delta^{AB}}{2(z-w)^2} + \frac{\mathrm{i}f_{ABC}\,J^C(w)}{z-w} + \dots. \tag{7}$$

The parameter k in the Kac–Moody algebra, called the 'level', is analogous to the parameter c in the conformal algebra. For a U(1) Kac–Moody algebra, $\hat{U}(1)$, it can be absorbed in the normalization of the current. However, for a non-abelian group G, it has an absolute meaning once the normalizations are carefully specified. Rather than giving general formulas, let me simply state that I choose $f_{ABC} = \epsilon_{ABC}$ in the case of SU(2). With this normalization convention, the algebra admits unitary representations if and only if k is a positive integer. Unitarity is essential for us, because we will use such algebras to define the positive-definite Hilbert space of physical states. An important formula, due to Sugawara (1986) and Sommerfield (1986), gives the energy–momentum tensor associated with an arbitrary Kac–Moody algebra:

$$T(z) = \frac{1}{k+\tilde{h}_G} \sum_{A=1}^{\dim G} :J^A(z)\,J^A(z):. \tag{8}$$

In the case of simply laced algebras ($G = A, D, E$) the dual Coxeter number $\tilde{h}_G$ equals c_A, where c_A is the quadratic Casimir number of the adjoint representation defined (with our normalization conventions) by

$$f_{ABC}f_{A'BC} = c_A\,\delta_{AA'}. \tag{9}$$

The associated central charge is $\quad c = k\dim G/(k+\tilde{h}_G). \tag{10}$

For example, in the case of $\widehat{\mathrm{SU}}(2)_k$, $\tilde{h} = 2$ and $c = 3k/(k+2)$.

An important restriction on the operator content of a conformal field theory is provided by the requirement of modular invariance (Cardy 1986). Specifically, it is necessary that the partition function

$$Z(\tau) = \sum N_{h_i \bar{h}_i} \chi_{h_i}(\tau) \, \chi^*_{\bar{h}_i}(\tau) \tag{11}$$

be invariant under the transformation group

$$\tau \to (a\tau + b)/(c\tau + d), \tag{12}$$

where a, b, c, d are integers satisfying $ad - bc = 1$. This group is generated by the transformations $T: \tau \to \tau + 1$ and $S: \tau \to -1/\tau$. The coefficients $N_{h_i \bar{h}_i}$ in (11) are non-negative integers representing the multiplicities of operators with conformal dimensions $(h_i, \bar{h}_i)$. The conformal characters $\chi_{h_i}(\tau)$ are the trace of $e^{2\pi i \tau(L_0 - \frac{1}{24}c)}$ in the module defined by the highest weight state $|h_i\rangle$ and its descendants. Clearly T invariance requires that the spins $h_i - \bar{h}_i$ be integers. Given a chiral algebra, such as the Kac–Moody algebras, it is in general a difficult problem to find all the possible modular invariant partition functions and to classify the associated conformal field theories. However, in the case of the Kac–Moody algebras $\widehat{SU}(2)_k$ the problem has been completely solved.

From the representation theory of $\widehat{SU}(2)_k$ one knows that the possible dimensions of primary fields are $l = 0, \frac{1}{2}, \ldots, \frac{1}{2}k$. Letting $\lambda = 2l + 1$ and $N = 2(k+2)$, the characters are (Kac & Peterson 1984)

$$\chi_\lambda(\tau) = \frac{1}{\eta^3(\tau)} \sum_{n=-\infty}^{\infty} (nN + \lambda) \exp\left[i\pi\tau(nN + \lambda)^2/N\right], \tag{13}$$

where $\eta(\tau)$ is the Dedekind eta function

$$\eta(\tau) = \exp\left(\tfrac{1}{12}\pi i\tau\right) \sum_{n=1}^{\infty} \left[-\exp\left(2\pi i n\tau\right)\right]. \tag{14}$$

Under an S transformation one has

$$\chi_\lambda(-1/\tau) = \sqrt{\left(\frac{2}{k+2}\right)} \sum_{\lambda'=1}^{k+1} \sin\left(\frac{\pi\lambda\lambda'}{k+2}\right) \chi_{\lambda'}(\tau).$$

By using the facts given above, the following theorem has been proved (Capelli et al. 1987; Gepner & Qiu 1987a): modular-invariant $\widehat{SU}(2)$ partition functions are in one-to-one correspondence with simply laced Lie algebras. In each case $k+2$ is the Coxeter number of the corresponding algebra. Moreover, the multiplicities of the diagonal terms are the Betti numbers of the corresponding algebras. The partition functions are listed in table 1. Because $\widehat{SU}(2)$ enters in a crucial way in the construction of various other conformal field theories, this result is directly relevant for them, as well.

TABLE 1. MODULAR-INVARIANT PARTITION FUNCTIONS FOR $SU(2)_k$

(Capelli et al. 1987.)

A_{k+1}	$\displaystyle\sum_{\lambda=1}^{k+1} |\chi_\lambda|^2$	$k \geqslant 1$
$D_{2\rho+2}$	$\displaystyle\sum_{\lambda_{\mathrm{odd}}=1}^{2\rho-1} |\chi_\lambda + \chi_{4\rho+2-\lambda}|^2 + 2|\chi_{2\rho+1}|^2$	$k = 4\rho, \ \rho \geqslant 1$
$D_{2\rho+1}$	$\displaystyle\sum_{\lambda_{\mathrm{odd}}=1}^{4\rho-1} |\chi_\lambda|^2 + |\chi_{2\rho}|^2 + \sum_{\lambda_{\mathrm{even}}=2}^{2\rho-2} (\chi_\lambda \chi^*_{4\rho-\lambda} + \text{c.c.})$	$k = 4\rho - 2, \ \rho \geqslant 2$
E_6	$|\chi_1 + \chi_7|^2 + |\chi_4 + \chi_8|^2 + |\chi_5 + \chi_{11}|^2$	$k = 10$
E_7	$|\chi_1 + \chi_{17}|^2 + |\chi_5 + \chi_{13}|^2 + |\chi_7 + \chi_{11}|^2 + |\chi_9|^2 + [(\chi_3 + \chi_{15})\chi^*_9 + \text{c.c.}]$	$k = 16$
E_8	$|\chi_1 + \chi_{11}|^2 + |\chi_{19} + \chi_{29}|^2 + |\chi_7 + \chi_{13} + \chi_{17} + \chi_{23}|^2$	$k = 28$

[44]

An important first step towards finding the general result for arbitrary Kac–Moody algebras has been taken by Gepner ($1987a$). He gives a complete list of modular-invariant partition functions without the restriction $N_{h_i \bar{h}_i} \geqslant 0$. In the form given, it is a very non-trivial problem to determine the subset that satisfies this restriction.

$$N = 2 \text{ SUPERCONFORMAL SYMMETRY}$$

By definition, an N-extended superconformal symmetry algebra is one containing N dimension-$\frac{3}{2}$ fermionic generators $G^\alpha(z); \alpha = 1, 2, \ldots, N$. It is also required that the OPE $G^\alpha(z) G^\beta(w)$ contain the term $2T(w) \delta^{\alpha\beta}/(z-w)$, where $T(w)$ is energy–momentum tensor. The $N = 2$ algebra, in addition to T, G^1, G^2, also contains a dimension-one current $J(z)$; J can be regarded as defining an abelian Kac–Moody algebra $\hat{U}(1)$. The $N = 2$ superconformal algebra is given by

$$
\left.
\begin{aligned}
T(z)\, T(w) &\sim \frac{c}{2(z-w)^4} + \frac{2T(w)}{(z-w)^2} + \frac{\partial T(w)}{z-w} + \cdots, \\[2mm]
G^\alpha(z)\, G^\beta(w) &\sim \left(\frac{2c}{3(z-w)^3} + \frac{2T(w)}{z-w} \right) \delta^{\alpha\beta} + i\left(\frac{2J(w)}{(z-w)^2} + \frac{\partial J(w)}{z-w} \right) \epsilon^{\alpha\beta} + \cdots, \\[2mm]
T(z)\, G^\alpha(w) &\sim \frac{3}{2} \frac{G^\alpha(w)}{(z-w)^2} + \frac{\partial G^\alpha(w)}{z-w} + \cdots, \\[2mm]
T(z)\, J(w) &\sim \frac{J(w)}{(z-w)^2} + \frac{\partial J(w)}{z-w} + \cdots, \\[2mm]
J(z)\, G^\alpha(w) &\sim i\epsilon^{\alpha\beta} \frac{G^\beta(w)}{z-w} + \cdots, \\[2mm]
J(z)\, J(w) &\sim \frac{1}{3} \frac{c}{(z-w)^2} + \cdots.
\end{aligned}
\right\}
\tag{15}
$$

The combinations $G^\pm(z) = (G^1(z) \pm iG^2(z))/\sqrt{2}$ are sometimes convenient to consider, as $G^+(z)\, G^+(w)$ and $G^-(z)\, G^-(w)$ are non-singular, whereas

$$
G^+(z)\, G^-(w) \sim \frac{2c}{3(z-w)^3} + \frac{2J(w)}{(z-w)^2} + \frac{2T(w) + \partial J(w)}{z-w} + \cdots.
\tag{16}
$$

In the case of the $N = 1$ superconformal algebra we distinguish two sectors, NS and R, according to whether $G(z)$ has half-integral or integral modes. In the case of the $N = 2$ algebra there are three distinct sectors that can be distinguished (Boucher *et al.* 1986):

(1) the R sector in which all operators have integer modes;

(2) the NS sector in which $G^1(z)$ and $G^2(z)$ have half-integral modes and $T(z)$ and $J(z)$ have integral modes;

(3) the T sector in which $G^1(z)$ and $J(z)$ have half-integral modes and $G^2(z)$ and $T(z)$ have integral modes. The T sector does not seem to play a role in applications to string theory.

When $J(z)$ has integer modes, namely in the R and NS sectors, it has a zero mode J_0, which is a U(1) charge. In these sectors, we can classify highest-weight states (and hence primary fields) not only by the eigenvalue h of the operator L_0, but also by the eigenvalue q of J_0. In the twisted sector $J(z)$ has no zero mode and hence there is no such charge to be defined.

To explicitly construct a large class of $N = 2$ models it is convenient to consider super-

Kac–Moody algebras. They contain dimension-$\frac{1}{2}$ fermionic currents $j^A(z)$ as superpartners of the usual dimension-1 bosonic currents $J^A(z)$. Because the $j^A(z)$ belong to the adjoint representation,

$$J^A(z)\,j^B(w) \sim \frac{\mathrm{i} f_{ABC}\, j^C(w)}{z-w} + \dots . \tag{17}$$

They have free fermion commutation relations, so choosing a convenient normalization,

$$j^A(z)\,j^B(w) \sim \frac{k\delta^{AB}}{2(z-w)} + \dots . \tag{18}$$

The unitary representation of a super-Kac–Moody algebra with the lowest possible level is given by setting the bosonic currents $J^A(z)$ equal to the fermion bilinears

$$J_f^A(z) = -(\mathrm{i}/k)\, f^{ABC} j^B(z)\, j^C(z). \tag{19}$$

In this case it is easy to see that the conformal anomaly is just that of $\dim G$ free fermions $c_f = \dim \frac{1}{2} G$, and the level of the Kac–Moody algebra is $k_f = c_A(G)$.

By using this special representation, the general representation of a super-Kac–Moody algebra can be obtained (Kac & Todorov 1985). Letting

$$J^A(z) = J_f^A(z) + \hat{J}^A(z), \tag{20}$$

one sees that $\hat{J}^A(z)$ defines a Kac–Moody algebra that is independent of the fermion fields. In other words, $\hat{J}^A(z)\, j^B(w)$ and $\hat{J}^A(z)\, J_f^B(w)$ are non-singular. If we therefore consider a level $\hat{k}$ representation of the algebra given by the $\hat{J}^A(z)$, we obtain a representation of the $J^A(z)$ algebra with level

$$k = \hat{k} + c_A(G) = \hat{k} + \tilde{h}_G \tag{21}$$

and central charge

$$c_G = \hat{k}\dim G/(\hat{k} + \tilde{h}_G) + \tfrac{1}{2}\dim G. \tag{22}$$

The allowed values of $\hat{k}$ are $0, 1, 2, \dots$, where the choice $\hat{k} = 0$ implies setting $\hat{J}^A = 0$. For simplicity it is assumed that G is simply laced so that $c_A(G) = \tilde{h}_G$, where $c_A(G)$ is the Casimir number defined in (9). The corresponding energy–momentum tensor associated with the group G is

$$T_G(z) = (1/k)\,(:\hat{J}^A(z)\,\hat{J}^A(z):\, - :j^A(z)\,\partial j^A(z):), \tag{23}$$

and the supercurrent is

$$G_G(z) = (2/k)\,(j^A(z)\,\hat{J}^A(z) - \tfrac{1}{3}(\mathrm{i}/k) f_{ABC}\, j^A(z)\, j^B(z)\, j^C(z)). \tag{24}$$

It is now easy to generalize the GKO construction (Goddard *et al.* 1985) to super-Kac–Moody algebras (Kazama & Suzuki 1988*a*). Let $J^a(z), a = 1, \dots, \dim H$, denote the generators of $\hat{H}$ as before. Then, just as for $\hat{G}$, we can write

$$J^a(z) = \tilde{J}^a(z) - (\mathrm{i}/k) f_{abc}\, j^b(z)\, j^c(z), \tag{25}$$

where f_{abc} are the H structure constants and $j^a(z)$ are $\dim H$ of the free fermions. The remaining $\dim G - \dim H$ free fermions are denoted $j^{\bar{a}}(z)$. The H supercurrent is

$$G_H(z) = (2/k)\,(j^a(z)\,\tilde{J}^a(z) - \tfrac{1}{3}(\mathrm{i}/k) f_{abc}\, j^a(z)\, j^b(z)\, j^c(z)), \tag{26}$$

and the coset model is defined by the difference

$$G(z) = G_G(z) - G_H(z) = (2/k)\,(j^{\bar{a}}(z)\,\hat{J}^{\bar{a}}(z) - \tfrac{1}{3}(\mathrm{i}/k) f_{\bar{a}\bar{b}\bar{c}}\, j^{\bar{a}}(z)\, j^{\bar{b}}(z)\, j^{\bar{c}}(z)). \tag{27}$$

Just as in the ordinary GKO construction, one can easily show that this has non-singular OPEs with the H currents $\hat{J}^a(w)$ and $j^a(w)$. It is therefore guaranteed to define an $N = 1$ superconformal theory with central charge $c = c_G - c_H$, where

$$c_G = \tfrac{3}{2}\dim G - (\tilde{h}_G/k)\dim G \tag{28}$$

and similarly for c_H. The corresponding energy–momentum tensor is

$$T = \frac{1}{k}\left(\hat{J}^{\bar{a}}\hat{J}^{\bar{a}} - \frac{\hat{k}}{k}j^{\bar{a}}\,\partial j^{\bar{a}} + \frac{2\mathrm{i}}{k}f_{ab\bar{c}}\,\hat{J}^a j^{\bar{b}}j^{\bar{c}} - \frac{1}{k}f_{\bar{a}\bar{p}\bar{q}}f_{\bar{b}\bar{p}\bar{q}}j^{\bar{a}}\,\partial j^{\bar{b}} - \frac{1}{k^2}f_{\bar{a}\bar{b}\bar{c}}f_{\bar{a}\bar{d}\bar{e}}j^{\bar{b}}j^{\bar{c}}j^{\bar{d}}j^{\bar{e}} \right). \tag{29}$$

Let us examine the SU(2) case in more detail. In this case $\dim(G/H) = 2$ and $k = \hat{k}+2$, so that

$$c = 3\hat{k}/(\hat{k}+2), \tag{30}$$

providing a derivation of the $N = 2$ minimal models. The index $\bar{a}$ now takes the values 1 and 2, whereas the index a takes the value 3. Thus the equations for $T(z)$ and $G(z)$ become

$$T = \frac{1}{k}[(\hat{J}^1)^2 + (\hat{J}^2)^2] + \frac{4\mathrm{i}}{k^2}\hat{J}^3 j^1 j^2 - \frac{\hat{k}}{k^2}(j^1\,\partial j^1 + j^2\,\partial j^2) \tag{31}$$

and

$$G \equiv G^1 = (2/k)\,(j^1\hat{J}^1 + j^2\hat{J}^2). \tag{32}$$

The construction of the $N = 2$ representation is completed by the identifications

$$G^2(z) = (2/k)\,(j^1\hat{J}^2 - j^2\hat{J}^1) \tag{33}$$

and

$$J = (2/k)\,\hat{J}^3 + (2\mathrm{i}\hat{k}/k^2)j^1 j^2. \tag{34}$$

As a final check we can compute

$$J(z)\,J(w) \approx \tfrac{1}{3}c/(z-w)^2, \tag{35}$$

with $c = 3\hat{k}/(\hat{k}+2)$, as required. Not surprisingly, the modular-invariant partition functions of the minimal models have the same ADE classification that we presented for $\widehat{\mathrm{SU}}(2)$ models.

We have seen that it is possible to associate unitary representations of the $N = 1$ superconformal algebra with cosets by a generalized GKO construction. We saw that the theory admits a second supercurrent and thus furnishes an $N = 2$ model in the case of $\mathrm{SU}(2)/\mathrm{U}(1)$, corresponding to the $N = 2$ minimal models. A natural question is, What is the most general choice of G/H for which the $N = 1$ construction in fact defines an $N = 2$ algebra? This question has been examined by Kazama & Suzuki (1988). They find that there is a second supercurrent of the form

$$G^2 = (2/k)\,(h_{\bar{a}\bar{b}}j^{\bar{a}}\hat{J}^{\bar{b}} - \tfrac{1}{3}(\mathrm{i}/k)\,S_{\bar{a}\bar{b}\bar{c}}j^{\bar{a}}j^{\bar{b}}j^{\bar{c}}) \tag{36}$$

provided that the following conditions are satisfied:

(i) $h_{\bar{a}\bar{b}} = -h_{\bar{b}\bar{a}}$ and $h_{\bar{a}\bar{p}}h_{\bar{p}\bar{b}} = -\delta_{\bar{a}\bar{b}}$,

(ii) $h_{\bar{a}\bar{p}}f_{\bar{p}\bar{b}e} = h_{\bar{b}\bar{p}}f_{\bar{p}\bar{a}e}$,

(iii) $f_{\bar{a}\bar{b}\bar{c}} = h_{\bar{a}\bar{p}}h_{\bar{b}\bar{q}}f_{\bar{p}\bar{q}\bar{c}} + 2$ perms,

(iv) $S_{\bar{a}\bar{b}\bar{c}} = h_{\bar{a}\bar{p}}h_{\bar{b}\bar{q}}h_{\bar{c}\bar{r}}f_{\bar{p}\bar{q}\bar{r}}$.

Note that (i) implies that $h_{\bar{a}\bar{b}}$ is a complex structure for the coset manifold G/H.

To analyse the implications of these equations, let us define $\phi_e = h_{\bar{a}\bar{b}}f_{\bar{a}\bar{b}e}$ and consider $X_{cd} =$

$f_{cde}\,\phi_e$. As $f_{cd\bar{e}} = 0$, by the H group property, $X_{cd} = h_{\bar{a}\bar{b}}\,f_{\bar{a}\bar{b}E}\,f_{cdE}$. The Jacobi identity allows us to cycle the indices $\bar{a}, \bar{b}, c$ on the fs. Also, using the antisymmetry of $h_{\bar{a}\bar{b}}$, one obtains

$$X_{cd} = -2h_{\bar{a}\bar{b}}\,f_{c\bar{a}E}\,f_{\bar{b}dE} = -2h_{\bar{a}\bar{b}}\,f_{c\bar{a}\bar{e}}\,f_{\bar{b}d\bar{e}} = -2h_{\bar{a}\bar{e}}\,f_{c\bar{a}\bar{b}}\,f_{\bar{b}d\bar{e}}, \tag{37}$$

where the last step uses property (ii). Now interchanging the labels $\bar{b}$ and $\bar{e}$ we see that X_{cd} is equal to its negative and hence vanishes. It therefore follows that ϕ_e can only be non-zero if e corresponds to a U(1) factor in H. Thus we conclude that a necessary condition for $N = 2$ superconformal symmetry is that H contain a U(1) factor.

Let us now consider the special case of a symmetric space ($f_{\bar{a}\bar{b}\bar{c}} = 0$). Manipulations similar to those of the previous paragraph allow one to show that

$$X_{\bar{c}\bar{d}} = f_{\bar{c}\bar{d}e}\,\phi_e = c_A\,h_{\bar{c}\bar{d}}. \tag{38}$$

For hermitian symmetric spaces there is just one U(1) factor. Let $e = 0$ correspond to this U(1) factor, so that the only non-zero component of ϕ is ϕ_0. Then we see that $h_{\bar{a}\bar{b}} \propto f_{\bar{a}\bar{b}0}$, with a normalization determined by condition (i). Thus we are able to associate an $N = 2$ superconformal model to every hermitian symmetric space. These were classified by Cartan and are listed in the book of Helgason (1978). Table 2 lists the irreducible hermitian symmetric spaces (for compact G) and the associated $N = 2$ central charges. There are several cases that give the special value $c = 9$ irreducibly.

TABLE 2. HERMITIAN SYMMETRIC SPACES AND THE ASSOCIATED $N = 2$ CENTRAL CHARGES
(Kazama & Suzuki 1988.)

G/H	$c_{G/H}$
$SU(m+n)/SU(m) \times SU(n) \times U(1)$	$3\hat{k}mn/(\hat{k}+m+n)$
$SO(n+2)/SO(n) \times SO(2)$	$3\hat{k}n/(\hat{k}+n)\ n \geqslant 2$
for $n = 1, SO(3)/SO(2)$	$3\hat{k}/(\hat{k}+2)$
$SO(2n)/SU(n) \times U(1)$	$\frac{3}{2}\hat{k}n(n-1)/(\hat{k}+2n-2)$
$Sp(n)/SU(n) \times U(1)$	$\frac{3}{2}\hat{k}(n+1)/(\hat{k}+n+1)$
$E^6/SO(10) \times U(1)$	$48\hat{k}/(\hat{k}+12)$
$E_7/E_6 \times U(1)$	$81\hat{k}/(\hat{k}+18)$

(2, 2) COMPACTIFICATION

Gepner (1987c, 1988) has investigated superstring compactifications that give models with $(10-2n)$-dimensional Poincaré symmetry by describing internal degrees of freedom as a sum of $N = 2$ minimal models with

$$c = \sum_i \frac{3k_i}{k_i+2} = 3n. \tag{39}$$

These constructions can be carried out both for type II superstrings and for heterotic strings. In the latter case modular invariance of partition functions (and hence of loop amplitudes) can be implemented by formulating an analogue of embedding the spin connection in the gauge group. This involves using the same sum of minimal models for the left-movers. In the case of four dimensions ($n = 3$), the remaining $22-9 = 13$ units of c_L are contributed by the level-one Kac–Moody algebra $SO(10) \otimes \hat{E}_8$. (At level one the central charge is equal to the rank of the algebra.) The $\hat{SO}(10)$ symmetry actually becomes enlarged to E_6, so that one has $E_6 \times E_8$ gauge symmetry, as in the case of Calabi–Yau compactification. In general, there is some additional 'accidental' gauge symmetry as well.

Gepner has explored two examples in considerable detail. The first example is based on five copies of the $k = 3$ model. (Note that each one has $c = \frac{9}{5}$.) The second uses one copy of the $k = 1$ model and three copies of the $k = 16_E$ model $(1 + 3 \cdot \frac{8}{3} = 9)$. These are referred to succinctly as 3^5 and $1 \cdot 16_E^3$. The subscript E means that the E_7 exceptional affine invariant is used in the construction. He then gives overwhelming circumstantial evidence that these two models correspond to known Calabi–Yau spaces. The evidence includes a count of the numbers of generations and antigenerations as well as an analysis of the discrete symmetries.

One remarkable feature of Gepner's results is that the minimal models are exactly soluble, whereas Calabi–Yau spaces are very unwieldy in general. The construction is completely algebraic, and so it is not understood how geometric structure arises. If one could write a formula for the metric tensor or curvature tensor of the manifold in terms of the conformal field theory, one would have solved a mathematical problem that is usually assumed to be hopeless. Of course, this may not be possible. (Some insight into these matters will be provided later.) The models have extra $U(1)$ gauge symmetries beyond the expected $E_6 \otimes E_8$. There is one associated with each contributing minimal model, but one gets used up in extending $SO(10)$ to E_6. Thus the number of $U(1)$ factors is one less than the number of contributing minimal models (four for 3^5 and three for $1 \cdot 16_E^3$). In other examples the gauge symmetry can be extended even further.

Thus the minimal model constructions correspond to Calabi–Yau compactifications at special points of their moduli space where there is a large discrete symmetry group and enhanced gauge symmetry. One wonders whether this makes them too special to be of much interest, or whether these special features could make them physically preferred. Although it is not yet known how to calculate such things, it seems conceivable that non-perturbative effects could induce a potential that depends on the moduli in such a way that the theory would 'roll' to such special points. This would make them particularly good candidates for phenomenology. But even if this is not so, their study still seems to be a useful exercise, because they have so many realistic features. Also, many of these features only depend on the topology of the space and not on the particular choice of moduli. Of course, one is not restricted to minimal models only. If one were to use one of the Kazama–Suzuki models that gives $c = 9$ irreducibly, there would not be any extra 'accidental' gauge symmetry.

It is an interesting challenge to try to make a complete classification of $(2, 2)$ superconformal models with $c = 9$. This is a formidable task, although it does not look as hopeless now, as it did a few years ago. The first step is to examine what can be done with minimal models only. The equation

$$c = \sum_i \frac{3k_i}{k_i + 2} = 9 \tag{40}$$

has 168 solutions, which have been enumerated (Lynker & Schimmrigk 1988). However, there is additional freedom in choosing modular invariants of the contributing minimal models. Allowing arbitrary combinations of A, D, and E invariants gives 1176 possibilities. The 228 that only use A and E invariants have been tabulated and the number of generations and anti-generations of E_6 multiplets has been given in each case (Lütken & Ross 1988). There is some redundancy, however, because the E_6 and E_8 minimal models are reducible, as explained later. Many more models could also be formed by using the new $N = 2$ models and by modding out by discrete symmetries. Most cases correspond to Calabi–Yau spaces, but some are orbifolds (Eguchi *et al.* 1988). (There may even be examples with both interpretations!)

[49]

Recently, there has been progress in evaluating the Yukawa couplings in these models (Distler & Greene 1988). The $(27)^3$ coupling turns out to be given exactly by the lowest-order (large radius) field theory approximation (Candelas *et al.* 1985), whereas the $(\overline{27})^3$ is not. The knowledge of these couplings should make it possible to evaluate mass ratios and other quantities of physical interest. Then we can examine how close models of this type can come to agreeing with experiment.

Singularity theory classification of $N = 2$ superconformal models

New insights into the classification of $(2, 2)$ superconformal models have been obtained recently (Vafa & Warner 1988; Martinec 1988) by using mathematical methods of singularity theory (Arnold 1981) (also known as catastrophe theory). In particular, this work explains how to construct the Calabi–Yau spaces corresponding to all Gepner-type compactifications (Greene *et al.* 1988). This subject is new and developing fast. We will settle here for a brief description of some of the basic concepts.

Many interesting $N = 2, d = 2$ theories can be described by an action of the form

$$S = \int d^2x \, d^4\theta K(\Phi_i, \bar{\Phi}_i) + \left\{ \int d^2x \, d^2\theta W(\Phi_i) + \text{c.c.} \right\}, \tag{41}$$

In this expression the fields Φ_i are chiral $N = 2$ superfields, meaning that they are annihilated by certain supercovariant derivatives. In terms of physical states $|\Phi_i\rangle = \Phi_i|0\rangle$, this means that the NS sector descendants $G^a_{-\frac{1}{2}}|\Phi_i\rangle$ are null. The superpotential W is a holomorphic function of the Φ_i. Because of the presence of the 'F term' (the one containing W), these systems can be regarded as $N = 2$ Landau–Ginsburg models.

The main idea is to study the renormalization group flow under scale transformations of the two-dimensional metric $g \to \lambda^2 g$ as $\lambda \to \infty$. The kinetic term (known as the 'D term') contains only irrelevant operators, and thus W determines the fixed point of the RG flow. Specifically, at a fixed point the chiral fields scale according to

$$\left. \begin{aligned} \Phi_i &\to \lambda^{\omega_i}\Phi_i, \\ W(\lambda^{\omega_i}\Phi_i) &= \lambda W(\Phi_i). \end{aligned} \right\} \tag{42}$$

In this case, one says that W is quasi-homogeneous with weights ω_i. The fixed point describes a conformally invariant model in which the conformal dimensions of Φ_i are $(h_i, \bar{h}_i) = (\frac{1}{2}\omega_i, \frac{1}{2}\omega_i)$. The main result is that a quasi-homogeneous function W uniquely characterizes an $N = 2$ superconformal model up to field redefinitions (with a finite non-zero jacobian) and the addition of trivial quadratic terms in new fields. The analysis requires that Φ^n have a dimension n times that of Φ, which is valid as a consequence of the $N = 2$ algebra. By studying the scaling of the partition function, and comparing with the Weyl anomaly formula (Polyakov 1981), one can show that the central charge is $c = 6\beta$, where

$$\beta = \sum_i \left(\tfrac{1}{2} - \omega_i\right) \tag{43}$$

is called the 'singularity index' of W.

An important notion in singularity theory is 'modality'. Roughly speaking, this is the number of parameters that characterize the model. It is not quite the same thing as the number of physical moduli. Remarkably, the classification of modality $m = 0$ singularities precisely corresponds to the $N = 2$ minimal models! The same ADE classification discussed earlier was known previously in singularity theory, because there is a prescription for associating Dynkin-like diagrams to singularity types. The $m = 0$ classification is

$$\left.\begin{array}{llll}
A_n: & x^{n+1} & k = n-1, \\
D_n: & x^{n-1}+xy^2 & k = 2n-4, \\
E_6: & x^3+y^4 & k = 10, \\
E_7: & x^3+xy^3 & k = 16, \\
E_8: & x^3+y^5 & k = 28.
\end{array}\right\} \tag{44}$$

An obvious problem is to find the superpotentials that correspond to the various Kazama–Suzuki models.

The power of this approach is illustrated by the fact that there are three reducible minimal models, which are readily identified. Namely, as $A_m \otimes A_n$ corresponds to $x^{m+1}+y^{n+1}$,

$$\left.\begin{array}{l}
D_4 \equiv A_2 \otimes A_2, \\
E_6 \equiv A_2 \otimes A_3, \\
E_8 \equiv A_2 \otimes A_4.
\end{array}\right\} \tag{45}$$

The first of the correspondences is proved by noting that a linear change of variables allows $x^3 + xy^2$ to be expressed as the sum of two cubes. The discrete symmetries of the minimal models are also readily understood. For example, the Z_{k+2} symmetry of the A_{k+1} model is generated by $x \rightarrow \exp\left[2\pi i/(k+2)\right]x$.

Each of Gepner's models is characterized by a superpotential given by the appropriate sum of polynomials. For example, the 3^5 model corresponds to five A_4 models: $W = \sum_{i=i}^{5}\Phi_i^5$. The corresponding Calabi–Yau space is given by the hypersurface $W = 0$ in CP^4. This fact is derived (Greene *et al.* 1988) by making an appropriate change of variables in the path integral

$$\int d\Phi_1 \ldots d\Phi_5 \exp\left\{i\int d^2x\,d^2\theta W(\Phi_i)\right\}. \tag{46}$$

Only some of Gepner's models correspond to the fully classified complete-intersection Calabi–Yau (CICY) spaces given by polynomial constraints in products of complex projective spaces (Candelas *et al.* 1988*a, b*). The appropriate generalization that accommodates all of Gepner's models, suggested by the structure of quasi-homogeneous functions, involves 'weighted projective spaces'. The space $WCP^N_{k_1 \ldots k_{N+1}}$ is defined by the identification $[z_1, \ldots, z_{N+1}] \sim [\lambda^{k_1}z_1, \ldots, \lambda^{k_{N+1}}z_{N+1}]$. One subtlety is that these have non-trivial fixed-point sets. Remarkably, the condition for the vanishing of the first Chern class is simply that $c = 9$.

Concluding remarks

It was argued several years ago that string theory is a 'strongly coupled' theory (Dine & Seiberg 1985; Kaplanovsky 1985), meaning that (non-perturbative) quantum effects would be important. This could cause one to worry that the study of classical ground states is a waste of time, but I think that would be an overreaction. The three-generation models clearly come quite close to giving the desired phenomenology; it would be foolish not to explore how far they can be pushed. Many realistic features emerge quite naturally: gravity, popular gauge groups and representations, chiral families of fermions, axions, symmetry-breaking mechanisms, supersymmetry, etc. (I do not understand how those who say 'There is not a shred of experimental evidence for string theory' can ignore all these successes.) It would be a very strange coincidence if classical solutions were completely off the mark.

Certainly, non-perturbative phenomena will be crucial for a complete understanding. It is quite clear that the $N = 1, D = 4$ supersymmetry of the Calabi–Yau or orbifold solutions is not broken at any order in perturbation theory. Also, the dilaton and other massless states do not acquire a mass. I think it is reasonable to expect these things to happen in the complete non-perturbative quantum theory, however. An encouraging result is the recent demonstration that the string perturbation expansion diverges (Gross & Periwal 1988). This suggests that not every classical solution need correspond to a quantum ground state, but that whenever one does the needed symmetry breaking and mass generation could occur. Also, there probably are instantons that give quantum tunnelling between different classical vacua. These types of effects might lift the enormous degeneracy with which we are currently faced.

One question that has received much attention in recent years is why the cosmological constant is so small (less than 10^{-120} in Planck units). I don't know the answer, but let me offer the following comments. Perhaps, as we have discussed, the correct string theory ground state is reasonably approximated by a classical solution with $N = 1$ supersymmetry in four dimensions. Because supersymmetry is unbroken at every order in the loop expansion, the cosmological constant undoubtedly vanishes at every order. The mystery that then needs to be understood is why the non-perturbative effects that break supersymmetry do not generate a cosmological constant at the same time. It seems to me that the problem must be addressed in the context of the complete theory and is very unlikely to be resolved by considerations that are not sensitive to Planck-scale physics. I have assumed that low-energy supersymmetry is required to solve the 'hierarchy problem'. This was the principle motivation for looking for supersymmetric solutions, although they do seem to fit in rather naturally. Still, it would be very reassuring and helpful to have supporting experimental evidence. Could a 'string miracle' other than space-time supersymmetry do the job? After all, we need one to eliminate the cosmological constant.

In conclusion, it would probably be foolhardy to predict dramatic phenomenological successes for string theory in the near term. Still, there are some encouraging possibilities that deserve to be pursued. We might get lucky!

I am grateful to L. Dixon, D. Gepner, W. Lerche, E. Raiten and G. Rivlis for helpful discussions. This work was supported in part by the U.S. Department of Energy under Contract no. DE-AC0381-ER40050.

REFERENCES

Arnold, V. I. 1981 *Singularity theory.* London math. lect. notes ser. **53**. Cambridge University Press.
Banks, T. 1987 TASI Lectures SCIPP 87/111.
Belavin, A. A., Polyakov, A. M. & Zamolodchikov, A. B. 1984 *Nucl. Phys.* B **241**, 333–380.
Boucher, W., Friedan, D. & Kent, A. 1986 *Phys. Lett.* B **172**, 316–322.
Candelas, P., Dale, A. M., Lütken, C. A. & Schimmrigk, R. 1988*a Nucl. Phys.* B **298**, 493–526.
Candelas, P., Horowitz, G., Strominger, A. & Witten, E. 1985 *Nucl. Phys.* B **258**, 46–74.
Candelas, P., Lütken, C. A. & Schimmrigk, R. 1988*b Nucl. Phys.* B **306**, 113–136.
Capelli, A., Itzykson, C. & Zuber, J.-B. 1987 *Nucl. Phys.* B **280** [FS 18], 445–465.
Cardy, J. L. 1986 *Nucl. Phys.* B **270** [FS 16], 186–204.
Dine, M. & Seiberg, N. 1985 *Phys. Rev. Lett.* **55**, 366–369; *Phys. Lett.* B **162**, 299.

Distler, J. & Greene, B. 1988 Cornell and Harvard preprint.
Eguchi, T., Ooguri, H., Taormina, A. & Yang, S. 1988 University of Tokyo preprint UT-536.
Gepner, D. 1987*a* *Nucl. Phys.* B **290** [FS 20], 10–24
Gepner, D. 1987*b* *Phys. Lett.* B **199**, 380–385.
Gepner, D. 1988 *Nucl. Phys.* B **296**, 757–778.
Gepner, D. & Qiu, Z. 1987 *Nucl. Phys.* B **285** [FS 19], 423–453.
Ginsparg, P. 1988 Les Houches Lectures HUTP-88/A054.
Goddard, P., Kent, A. & Olive, D. 1985 *Phys. Lett.* B **152**, 88–92.
Goddard, P., Kent, A. & Olive, D. 1986 *Communs math. Phys.* **103**, 105–119.
Goddard, P., Nahm, W. & Olive, D. 1985 *Phys. Lett.* B **160**, 111–116.
Goddard, P. & Olive, D. 1986 *Int. J. mod. Phys.* A **1**, 303–414.
Green, M. B., Schwarz, J. H. & Witten, E. 1987 *Superstring theory* (2 vols). Cambridge University Press.
Greene, B., Vafa, C. & Warner, N. 1988 Preprint HUTP-88/A047.
Gross, D. & Periwal, V. 1988 *Phys. Rev. Lett.* **60**, 2105–2108.
Helgason, S. 1978 *Differential geometry, Lie groups, and symmetric spaces*. Academic Press.
Kac, V. G. 1985 *Infinite dimensional Lie algebras*. Cambridge University Press.
Kac, V. G. & Peterson, D. 1984 *Adv. Math.* **53**, 125–264.
Kac, V. G. & Todorov, I. T. 1985 *Communs math. Phys.* **102**, 337–347.
Kaplunovsky, V. 1985 *Phys. Rev. Lett.* **55**, 1036–1038.
Kazama, Y. & Suzuki, H. 1988 University of Tokyo preprint UT-Komaba 88-8 (revised); University of Tokyo preprint UT-Komaba-88-12.
Lerche, W., Vafa, C. & Warner, N. 1989 Harvard preprint HUTP-88/A065.
Lütken, C. A. & Ross, G. G. 1988 *Phys. Lett.* B **213**, 152–159.
Lynker, M. & Schimmrigk, R. 1988 Texas preprint UTTG-05-88.
Martinec, E. 1988 Preprint EFI 88-76.
Peskin, M. 1987 *1986 TASI lectures* (ed. H. Haber). World Scientific.
Polyakov, A. M. 1981 *Phys. Lett.* B **103**, 207–210, 211–213.
Sommerfield, C. 1968 *Phys. Rev.* **176**, 2019–2025.
Sugawara, H. 1968 *Phys. Rev.* **170**, 1659–1662.
Vafa, C. & Warner, N. 1988 Preprint HUTP-88/A037.

Discussion

J. R. Ellis, F.R.S. (*Theory Division, CERN, Geneva, Switzerland*). Professor Schwarz has discussed extensively $(2,2)$ compactifications. As he knows, more general compactifications are compatible with $N = 1$ space-time supersymmetry. Have any advances been made recently in the classification of such $(2,1)$ and $(2,0)$ compactifications?

J. H. Schwarz. As Dr Ellis probably knows, $(2,0)$ and $(2,1)$ models have been discussed in recent papers by Dine & Seiberg (1985), Distler & Greene (1988), and Cvetic. I am not an expert in these matters, but I am not aware of any recent advances indicating whether such schemes could be realistic.

Phil. Trans. R. Soc. Lond. A **329**, 373–393 (1989)

Printed in Great Britain

The (low-energy) physics of the superstring

By G. G. Ross

Department of Theoretical Physics, University of Oxford, 1 Keble Road, Oxford OX1 3NP, U.K.

Superstring theories offer the prospect of a finite unified theory of the four fundamental interactions. However, their implications for low-energy physics are difficult to determine because there must be several stages of symmetry breaking leading from the underlying multidimensional supersymmetric string theory, relevant at the scale of string structure (not more than $O(10^{-32}$ cm$)$), to the four-dimensional non-supersymmetric effective field theory relevant for physics at scales probed in the laboratory (not less than $O(10^{-15}$ cm$)$). I discuss these stages of compactification, supersymmetry breaking and gauge-symmetry breaking and show how the resulting low-energy theory may closely approximate the observed structure. The implications for the multiplet structure, the masses and the couplings of the observed particle states are considered.

1. Introduction

String theories provide the basis for a finite unification of the strong, the electromagnetic, the weak and the gravitational interations; perhaps they do indeed represent the ultimate 'theory of everything'. However, direct tests for underlying string structure are likely to be very difficult for the string structure is evident only at very short distance scales of order M_{P}^{-1}, the inverse Planck scales, i.e. $O(10^{-32}$ cm$)$. The physics at such short distances may have been probed in the very early universe and it is possible some observable relic of this period may exist but, so far as experimental tests in the laboratory are concerned, the 'stringy' aspects of particle states are negligible and tests of an underlying string theory must rely on indirect information such as the spectrum, the masses, the interactions and the couplings of the observed states. These states are essentially massless in Planck units and so laboratory tests concern the zero modes of the string; for this reason this paper is concerned with the physics of these zero modes.

The reason these tests provide only indirect evidence of the string substructure is that between the Planck scale, relevant for string physics, and the scale probed in the laboratory there must be several mass scales associated with further stages of symmetry breaking. There is the compactification scale needed to leave just four flat space-time dimensions. There may be scales associated with gauge-symmetry breaking to reduce the gauge group to that observed at low energies. There is the supersymmetry breaking scale needed to break the four-dimensional supersymmetry and, finally, there is the electroweak breaking scale needed to give mass to the W bosons the quarks and the leptons. These stages of symmetry breakdown serve to obscure the underlying string physics; much of the detailed low-energy structure following from the compactification stage. Although I believe the underlying string dynamics should determine this compactification, at present our analytical techniques do not allow us uniquely to determine the vacuum state of the string. As a result there is a large number of candidate string vacua with a corresponding uncertainty in the string predictions for low-energy physics. Accepting this limitation in our current understanding, it is still possible to make progress by analysing the most promising string vacua to see if their low-energy physics could be consistent with observation. For a given vacuum configuration the spectrum, masses and couplings should be determined so the comparison with experiment is likely to be very challenging. If a viable string model is shown to exist, the multifaceted problem of understanding the spectrum, masses and couplings of the observed states becomes one of trying to understand why a particular vacuum configuration is chosen.

Broadly, this paper falls into two parts. The first is fairly general, surveying the expectations of string theories for low-energy physics. The second part discusses how close specific attempts have come to a realistic low-energy theory. The discussion is restricted to the superstring, string theories with a world-sheet supersymmetry, because only in these theories is there the possibility of a residual supersymmetry in the four-dimensional world, as is needed to provide a solution to the mass hierarchy problem in a theory that does not have composite fields at a low energy-scale.

Briefly, the hierarchy problem (Gildener 1976; Gildener & Weinberg 1976; 't Hooft 1980) is the problem of maintaining a low scale of electroweak breaking in a theory with other much larger scales Λ (in string theories the compactification, M_c, or the Planck scale M_P). In the absence of a space-time supersymmetry there is nothing to prevent the Higgs scalar, Φ, from acquiring a mass term $M^2\Phi^\dagger\Phi$ invariant under all gauge and chiral symmetries. Even if at tree level $M^2 = 0$, radiative corrections induce a mass M at scale μ of $O(\alpha\Lambda^2\ln(\Lambda^2/\mu^2))$, where $\alpha = g^2/8\pi^2$ following from the one-loop graph with a virtual W exchanged between the Higgs scalar states (where g is the electroweak coupling). Thus the scale Λ^2 of new physics is constrained by $\Lambda^2 \lesssim M_W^2/\alpha$ if the Higgs mass, and hence the electroweak breaking scale, is to be of $O(M_W)$. In the standard model the sceptic might argue that the hierarchy problem is illusory for M^2 must be renormalized. In a string theory, however, one cannot evade the problem for the theory has no divergences and the contribution from the standard model will be cut off at a scale Λ, the compactification or the string scale. Hence the only consistent string theories (if they do not have composite Higgs scalars) must have a low-energy supersymmetry. For example, the W contribution to M^2 is cut off by the wino contribution, leaving a residual contribution as above with $\Lambda^2 = M_{\tilde{W}}^2$, the wino mass related to the supersymmetry breaking scale.

2. String theories

The essential feature of a string theory (Green *et al.* 1987*b*) that distinguishes it from a conventional field theory is that the states in the theory are not pointlike but, at some very short distance scale, have a distribution in a new dimension, σ. Thus string theories start with a world-sheet for the states in the theory, not a world-line. The world-sheet has both a space and a time dimension with coordinates σ and τ. The remaining index, μ, labels the 'target space' that describes the other degrees of freedom of the state, including the normal (four-dimensional) space-time degrees of freedom and further internal degrees of freedom (gauge, etc.). To define a viable theory, the following conditions are imposed.

1. *Modular invariance.* This is equivalent to reparametrization invariance on the world-sheet and requires that physics should not depend on how one parametrizes the underlying world-sheet with coordinates σ and τ.

2. *Conformal invariance.* The action of the string theory is defined in terms of the action on the two-dimensional world-sheet. There is an overall dimensionful coupling (the string tension) but the two-dimensional action itself does not involve dimensionful parameters and is required

to be conformally invariant, i.e. the underlying theory is a two-dimensional conformally invariant field theory.

3. *Absence of anomalies.* By making all states have an extension in the σ dimension, the infinities of conventional field theory associated with point interactions are ameliorated and, it is thought, string field theories give rise to finite theories of the fundamental interactions, including gravity. However, this is only possible if there are no anomalies in the theory, so an additional constraint must be added requiring that the theory be anomaly-free in the gauge and gravitational interactions. The absence of conformal anomalies places a strong constraint on the target space labelled by the μ index ($\mu = 1, ..., D$), namely that it should cancel the conformal ghost anomaly, which requires $D = 26$ for the bosonic string and 15 in a supersymmetric theory. As a result $D = 26$ or 10 for the non-supersymmetric and supersymmetric versions of the string respectively in D flat dimensions, because a free bosonic degree of freedom contributes 1 and a fermionic degree of freedom contributes $\frac{1}{2}$ respectively to the conformal anomaly. There is a intimate relation between the absence of anomalies and the modular-invariance constraint discussed above. The local conformal invariance that is achieved once the conformal anomaly is eliminated by the choice of target space corresponds to local reparametrization or modular invariance. It turns out that the requirement of global modular invariance is a necessary and sufficient condition for the elimination of all remaining gravitational and gauge anomalies.

4. *World-sheet supersymmetry.* Although not a necessary requirement for a string theory, I also demand that there be a $N = 1$ supersymmetry in the string theory compactified to $d = 4$ flat space-time dimensions. This allows for a solution of the gauge hierarchy problem so that a natural separation of the scale, $M_{\rm W}$, of electroweak breaking from the string scale $(\alpha')^{-1} = O(M_{\rm P})$ is possible. As discussed above, the finiteness of the string theory makes the gauge hierarchy problem *more* pressing, and I see no prospect of a viable low-energy theory being built from a non-supersymmetric string theory unless it corresponds to a theory in which light Higgs bosons are composite fields.

3. String compactification

The original realizations of the string conditions listed in §2 identified the D target-space indices with flat space-time, leading to the bosonic string in $D = 26$ dimensions, the fermionic string in $D = 15$ dimensions and the heterotic string with left-moving states described by the bosonic string in $D = 26$ dimensions and the right-moving states described by the fermionic string in $D = 15$ dimensions. As a result these models must undergo a stage of compactification in which all but four of these dimensions are associated with a compact space. The string states may freely propagate in the four flat dimensions, but will be confined to a distance R characterizing the dimension of the compact space in the compact dimension. As a result until distance scales of order R are probed ($R \sim M_{\rm c}^{-1} \sim 10^{-31}$ cm) the effective theory will appear to have four space-time dimensions. A characteristic feature of a compactified theory is the quantization of the level structure; the imposition of compact boundary conditions causes the levels in split with mass splitting proportional to R^{-1}. As a result a given compactification will make a prediction for the spectrum of states and the comparison of this spectrum for the zero-modes with that observed at low energies, the quarks, the leptons, the gluons, the photons and the W bosons, provides a stringent test of the compactification.

There has been considerable theoretical effort directed towards the analysis of compactifications consistent with the underlying string dynamics and a large number of possibilities have emerged (J. H. Schwarz, this Symposium).

3.1. *Calabi–Yau compactification*

For the fermionic string the condition the 10-dimensional space should become $M \times K$, where M is four-dimensional Minkowski space and K a six-dimensional compact manifold, together with the condition there should be a residual $N = 1$ supersymmetry led to the conclusion (Candelas *et al.* 1985) that K should have SU(3) holonomy, i.e. a Calabi–Yau space. In the heterotic string the most natural generalization is to identify the spin connection with the gauge connection coming from an SU(3) subgroup of the $E_8 \times E_8'$ gauge group coming from the bosonic sector. As a result the gauge group after compactification is $E_8 \times E_6$. The gauge non-singlet chiral superfields in the theory all originate in the original $E_8 \times E_8'$ gauge bosons. After compactification the E_8' gauge bosons have a harmonic expansion of the form

$$A_\mu^{\tilde{\alpha}}(x, y) = \sum_i^N X_i^\alpha(x) \, Y_{ab}^i(y), \tag{3.1}$$

where we have written the E_8' index $\tilde{\alpha}$ in terms of an E_6 index α and an SU(3) (gauge connection) index b by using the decomposition $E_8 \supset E_6 \times SU(3)$. The index a is an SU(3) space-group index, where we have constructed three complex coordinates a from the six internal coordinates, μ.

Now the E_8 adjoint representation is 248-dimensional, and under $E_6 \times SU(3)$ it decomposes as

$$(248) = (78, 1) + (27, 3) + (\overline{27}, \overline{3}) + (1, 8). \tag{3.2}$$

Consider the representation $(\alpha, b) \equiv (27, 3)$. In this case b is a triplet index and $Y_{iab}(y)$ transforms as a triplet under both space and internal symmetry groups. Because we have identified the spin and the gauge connections, these indices are both SU(3) holonomy indices; Y_{ab} has mixed symmetry under the interchange of a and b, but we may identify it with a differential form by rewriting $Y_{ab}(y) = \epsilon_{acd} Y_b^{cd}(y)$. From, for example, equation (3.1) we see the zero modes in four dimensions $X_i(x)$ are associated with zero-modes $Y^i(y)$ in the six-dimensional compact spaces. The number of such fields is given by the number $h_{2,1}$ of related $(2, 1)$ forms $Y \equiv Y_b^{cd} \, dz_A^b \, d\bar{z}_{cA} \, d\bar{z}_d$. Thus one establishes a connection between the Hodge numbers $h_{i,j}$ of the compact manifold and the four-dimensional massless states (Hübsch 1987; see Candelas (1987) for a general review). There are $h_{2,1}$ chiral superfields transforming as the 27 of E_6, $h_{1,1}$ chiral superfields transforming as the $\overline{27}$ of E_6 and a number of gauge single chiral superfields not completely determined by the Hodge numbers but at least the $(h_{2,1} + h_{1,1})$ moduli of the space.

How do these expectations fit with observation characterized by the 'standard model'?

The gauge group is $SU(3) \times SU(2) \times U(1)$, which neatly fits into the $SU(3) \times SU(3) \times SU(3)$ maximal subgroup of E_6. This assignment also fits well for the matter multiplets, for the 27-dimensional representation of E_6 has components transforming as $(1, 3, \overline{3}) \oplus (3, \overline{3}, 1) \oplus (\overline{3}, 1, 3)$ under this maximal subgroup and neatly accommodates a single family of leptons and quarks;

$$(1, 3, \overline{3}) = \begin{bmatrix} H_1^0 & H_2^+ & e^+ \\ H_1^- & H_2^0 & v_e \\ e^- & v_R & N \end{bmatrix}_R, \quad (3, \overline{3}, 1) = \begin{bmatrix} u \\ d \\ D \end{bmatrix}_L, \quad (\overline{3}, 1, 3) = \begin{bmatrix} u^c \\ d^c \\ D^c \end{bmatrix}_L. \tag{3.3}$$

Here there is room for the electron, e, its neutrino ν_e and the up and down quarks, u and d. There is also room for Higgs doublets H_1 and H_2. There are additional neutral lepton states ν_R and N and colour triplet states D.

The gauge group E_6 must be broken to give a viable theory and some of the states in equation (3.3) may be expected to be heavy.

However, the picture emerging is quite encouraging, at least the gauge group structure is large enough to accommodate the observed states. (In this context it may be worth recalling that the original extended supergravity models failed because their gauge group structure was just too small.) The family duplication may easily be understood in these models if the Hodge numbers predict more than one excess copy of $27 - \overline{27}$ representations (i.e. $h_{2,1} - h_{1,1}$ families). As I discuss in §4 there are only a very small number of Calabi–Yau models with just three families but their existence does support the feasibility of string models as far as their implications for the multiplet structure.

Finally, note that the E_8 sector has not been identified with any of the known states. This is because all known matter states are chiral; they do not belong to real representations and E_8 has only real representations. Attempts to build realistic models from the superstring make use of such a 'hidden' sector to provide a trigger for supersymmetry breaking (Ferrara *et al.* 1983; Derendinger *et al.* 1985; Dine *et al.* 1985*b*; Binétruy & Gaillard 1986; Breit *et al.* 1985*b*; Mangano 1985). It is thought, in analogy with our experience of quantum chromodynamics (QCD), that the E_8 group will be confining and all its states will be massive at the confinement scale. Moreover, the gaugino condensate that seems likely to form in this sector breaks supersymmetry and may act as the trigger for supersymmetry breaking in the 'visible' E_6 sector. Because the two sectors communicate only by gravitational strength, breaking effects are very small in the visible sector (Cremmer *et al.* 1983; Ellis *et al.* 1984*a, b*; Coughlan *et al.* 1988; Ross 1988*b*).

3.2. *Four-dimensional string theories*

Subsequent to the discovery of Calabi–Yau compactifications it has been realized that there is a much larger class of possibilities for compactified string models. This may be seen most directly by first observing that the definition of a string given in §2 does not require that all D dimensions be initially interpreted as flat space-time dimensions. Provided the conditions of modular invariance, etc., be realized some of these dimensions may, from the start, be compact or internal. This has led to the study of four string theories (J. H. Schwarz, this Symposium), the set of models satisfying the conditions of §2 with just four flat space-time dimensions.

3.2.1. *Orbifold models*

The simplest compactified models use flat or toroidal compactification to describe the internal dimensions. Thus, for example, six of the fermionic string dimensions may be identified with the real six torus $(T_R)^6$ defined by a $[SU(3)]^3$ lattice. This has an $N - 4$ supersymmetry in four dimensions, too large for a realistic theory, so to reduce this to an $N = 1$ supersymmetry the torus is modded out by a discrete point group P to form and 'orbifold' that breaks the $N = 4$ supersymmetry to an $N = 1$ supersymmetry introducing curvature at a discrete set of points (this is the discrete analogue of the $SU(3)$ holonomy group met in Calabi–Yau compactification where, however, the curvature is distributed throughout the manifold). For the simple example the point group may be chosen as Z_3, the discrete rotations of the three $SU(3)$ lattices simultaneously by 120°. The orbifold is known as the Z_3 orbifold (Dixon *et al.*

1985, 1986; Ibañez *et al.* 1987*a, b*; Chamseddine & Derendinger 1987). The action of P leaves $3 \times 3 \times 3 = 27$ fixed point, so the structure thus defined is an orbifold, not a manifold. In the orbifold there are two types of closed strings: 'untwisted' and 'twisted'. The untwisted string is one closed on the torus. The twisted strings rely on an action of the point group to close. The simplest heterotic versions of orbifold models use the same orbifold construction for left- and right-movers,

$$K = T_R^6/P \otimes T_L^6/P \otimes T_L^{16}/G, \tag{3.4}$$

where there may also be isometries G of the additional gauge degrees of freedom in the bosonic sector reducing the gauge group after compactification. These orbifold models have a discrete holonomy group, a subgroup of the SU(3) holonomy group used in Calabi–Yau compactification. As a result the gauge group structure (with trivial gauge isometries) is $E_8 \times E_6 \times G_0$ where G_0 is a subgroup of SU(3). For the case of abelian orbifolds, $G_0 = SU(3)$, $SU(2) \times U(1)$ or $U(1)^2$.

Orbifold models have been extensively studied, and it has been shown how to go beyond the left–right symmetry assumption that went into the Calabi–Yau construction to form asymmetric orbifold compactifications,

$$K = T_R^6/P_R \otimes T_L^6/P_L \otimes T_L^{16}/G. \tag{3.5}$$

It also proves possible to use the full 22 ($= 26 - 4$) internal dimensions of the bosonic string to provide gauge degrees of freedom leading to the possibility of a rank 22 gauge group (Narain 1986; Narain *et al.* 1986, 1987). Of course this is much too large to describe the gauge symmetries observed at low energies, and considerable effort has gone into finding 'realistic' orbifold models (Ibañez *et al.* 1987*a–c*, 1988; Bailin *et al.* 1987; Casas *et al.* 1987; Casas & Muñoz 1988). The best candidate models now have three generations of quarks and leptons and with non-trivial gauge isometries and a gauge group $SU(3) \times SU(2) \times U(1)^n$, where $n > 1$. Because the curvature is only at isolated points, the advantages of toroidal compactification are kept and all the properties of the model may be computed (Cvetic 1987).

3.2.2. *Fermionic formulation*

The formulation of string theories in spaces with isolated points of curvature has also been extensively explored using the 'fermionic' formulation in which the internal degrees of freedom are described by fermionic coordinates (Kawai *et al.* 1986; Lerche *et al.* 1987; Müller & Witten 1986; Antoniadis *et al.* 1987*a*; Schellekens 1987). Although partly overlapping the bosonic orbifolds formulation it also gives rise to new compactification not easily related to the orbifold construction. Once again it has been shown possible to build three-generation models, as I discuss in the next section.

3.2.3. $N = 2$ *discrete series*

A recent development seeks to construct the string theory in four dimensions by using tensor products of $N = 2$ unitary discrete series chosen to cancel the conformal anomaly (Gepner 1988). This technique provides a generalization of the orbifold and fermionic methods allowing for the construction of supersymmetric conformal field theories with curvature.

Because the $N = 2$ discrete series is completely solvable they retain the advantages of the other formulations in that the partition function is completely known at string tree level. Remarkably, these models generate some complete intersection Calabi–Yau (CICY) theories allowing for the complete computation in these theories of the spectrum and couplings.

The $N = 2$ minimal models are labelled by a positive integer, k, and the kth models has a construction to the central charge (i.e. the contribution to the conformal anomaly) given by $c = 3k/(k+2)$.

4. Low-energy implications of string compactifications

In this section I briefly review the characteristics to be expected for the low-energy physics following from string compactifications. Because of the large number of candidate string vacua it is not possible at present to make definitive statements, but certain features seem to be common to a wide class of realistic compactification schemes; in the next section I illustrate how these generic features are realized in specific models.

4.1. *Vacuum structure*

In addition to the problem of the large number of disconnected string compactifications within a particular scheme there may also be several undetermined parameters. The first set of these correspond to the moduli, which determine the shape and size of the manifold. In Calabi–Yau models there are $(h_{2,1} + h_{1,1})$ such moduli that correspond to chiral supermultiplets whose scalar components have undetermined vacuum expectation values (vevs) because their potential vanishes identically. However, the symmetries of the model may be such that the potential of many other fields also vanishes identically, or to a high degree of accuracy, giving rise to further 'flat' directions in the potential. It is along such flat directions that gauge non-singlet fields may acquire large vevs reducing the gauge symmetry.

The determination of these vevs occurs only once the vacuum degeneracy along the flat direction is lifted. This happens once the low-energy $N = 1$ supersymmetry is broken as is expected to happen via the formation of condensates in the hidden sector. In supergravity theories such a gaugino condensate breaks supersymmetry (Ferrara *et al.* 1983; Derendinger *et al.* 1985; Dine *et al.* 1985 *b*; Binétruy & Gaillard 1986; Breit *et al.* 1985 *b*; Mangano 1985) giving rise to a gravitino mass and supersymmetry breaking in the hidden sector. These effects are communicated by gravitational-strength couplings to the visible sector giving masses to the scalar components of chiral superfields and the fermion component of gauge supermultiplets (Cremmer *et al.* 1983; Ellis *et al.* 1984 *a, b*; Coughlan *et al.* 1988; Ross 1988 *b*). Because of large radiative corrections involving the light states in the theory some of the squared scalar masses are driven negative triggering spontaneous symmetry breakdown as the scalar acquires a vev at the minimum of the potential (Inoue *et al.* 1982, 1984; Yamagishi 1983; Alvarez-Gaume *et al.* 1983; Ibañez *et al.* 1985; Gato *et al.* 1985). Detailed studies (Ross 1988) of these effects show that they do indeed fix the moduli, often at values enhancing the residual discrete symmetries of the model. It seems likely, too, that gauge symmetries will be broken at scales close to the compactification scale whenever there are flat directions involving gauge-non-singlet fields. Thus a large part of the vacuum degeneracy is lifted via supersymmetry breaking effects, going some way towards answering the big question of why a particular vacuum configuration is chosen. This applies to the set of string vacua continuously connected via change of moduli or other vevs, for the small (on the compactification scale) supersymmetry breaking effects may still cause these parameters to change to reduce the vacuum energy. However, different string vacua not continuously connected will not be driven to a unique ground state by these small supersymmetry breaking effects because the potential energy barrier between these vacua is of the order of the compactification scale. The question of how the choice between such vacua is made is a question of how the universe evolved from early times at which it had a temperature of the order of the compactification scale. Thermal effects do not respect supersymmetry because of the different statistics obeyed by fermions and bosons, so we may expect these effects to split the string vacua degeneracy and consequently a study of string vacua at high temperature may show how the true vacuum state is chosen from the set separated by large potential barriers.

4.2. *Gauge and discrete symmetries*

As has already been stressed the gauge symmetry is determined by the compactification and the only definite constraint is that it should have rank 22 or less with a supersymmetry $N = 4$ or less. Specific vacua have much smaller symmetry groups and here I concentrate on models that have a gauge symmetry given by E_6 or smaller together with a hidden-sector gauge group that plays no role in the low-energy structure apart from providing a trigger for supersymmetry breaking needed to solve the gauge hierarchy problem. Although it seems attractive to have a compactification scheme that has just the standard $SU(3) \times SU(2) \times U(1)$ gauge group this is by no means essential for there may be stages of gauge symmetry breaking by the usual Higgs mechanism after compactification in which scalar fields acquire vacuum expectation values. Whether this happens is determined by the form of the scalar potential following from the compactification, and the occurrence of flat directions involving gauge non-singlet fields. The existence of such flat directions seems to be a generic feature in string compactifications and so the general expectation for the low-energy gauge group is that it should be much smaller than the maximal rank 44 gauge group. As is seen in specific models it is easy to reduce the gauge symmetry to just that of the standard model at a high scale, the question becomes one of why there is *any* residual gauge symmetry. The answers to this relies on whether the flat directions are 'exhausted' before all gauge symmetries are broken; specific models may give rise to additional light gauge bosons (Dine *et al.* 1985*a*; Breit *et al.* 1985*a*; del Aguila *et al.* 1986).

In part, the existence of flat directions is due to additional discrete symmetries present in the theory after compactification (Witten 1986; Greene *et al.* 1986). Their origin lies in the discrete symmetries of the compactification manifold or in the underlying two-dimensional conformal field theory and their trace in the low-energy theory may provide a good signal of the underlying physics. In specific models they may play an important role in suppressing baryon-number-violating decay processes and limiting the form of the light quark- and lepton-mass matrices (Greene *et al.* 1986, 1987).

4.3. *Matter content*

The standard model has matter content that comes in three copies or families of a distinctive multiplet pattern of quarks and leptons. The topological structure of the compactification scheme determines the matter content and thus may provide an explanation for this family duplication and the multiplet structure within a family. There may also be left light, new states not needed for the standard model. For example, each family in E_6 models is 27 dimensional and has an additional charge $Q = -\frac{2}{3}$ coloured state, a right-handed neutrino state, plus two charged lepton and five neutral lepton helicity states. These states may receive mass on the spontaneous breaking of the gauge symmetry, but there are often left light some residue of such new states.

4.4. *Couplings, masses and mixing angles*

Perhaps the most exciting prospect for testing string theories lies in the fact that the couplings of the model are, in principle, fixed on compactification. In practice the analytical techniques available may not allow all this information to be determined. In Calabi–Yau compactifications the metric is not known and so only limited information on couplings based on topology has been obtained. However, in constructions based on free fields the couplings are determined and this may sometimes be used to determine the couplings in other models. For example, orbifold calculations may be extended to the associated Calabi–Yau models related via blow-up of the orbifold singularities. The recent models based on tensor products of minimal $N = 2$ theories may, more directly, determine the couplings in Calabi–Yau models at particular points in moduli space (Distler & Greene 1988; Gepner 1987). These predictions do not rely on discrete topological invariants and so, unlike the predictions for spectra, they may be sensitive to corrections such as non-perturbative effects, higher loop corrections, etc. The question of the stability of coupling predictions has received much attention and it has been shown it is possible to obtain reliable predictions for some of the couplings in the theory responsible for mass generation in the standard model (Distler & Greene 1988; Lutken & Ross 1988). In particular, using the world-sheet conformal symmetry (Dixon 1987) one may show that the $(27)^3$ couplings do not receive non-perturbative world-sheet instanton corrections and so their calculation in the conformal field theory limit that corresponds to a particular compactification radius may be used for the Calabi–Yau theory at arbitrary radius. The couplings, $(\overline{27})^3$, do, however, receive large non-perturbative corrections and so the conformal field theory estimate does not apply at arbitrary radius. However, these couplings are determined via topological techniques in the field theory limit and the non-perturbative corrections about this limit are likely to be small and in some cases can be calculated. (Similar considerations may be applied to the other trilinear couplings of the model.) Beyond string tree level few calculations have been made but, in theories with a residual supersymmetry, the non-renormalization theorems ensure that these corrections to couplings are small.

Being an effective field theory below the compactification scale, there are also non-renormalizable couplings to be considered. It has been shown that these only occur in non-perturbative order (Cvetic 1987; Burgess *et al.* 1986) and so, if the compactification scale is below the Planck scale (as is expected if the hierarchy problem is to be solved), these terms will be small. They may, however, still be important in theories with large-scale vevs developing along flat directions; in particular, neutrino masses are quite sensitive to non-renormalizable terms (Greene *et al.* 1987; Nandi & Sarkar 1986). At present there are few explicit calculations of non-renormalizable couplings, but the possible terms are strongly constrained by the gauge and discrete symmetries of the compactification scheme. Given a knowledge of the couplings of the theory it is, in principle, possible to determine all of the properties of the low-energy theory. In particular, after electroweak breaking the cubic couplings will determine the quark and lepton mass matrices and hence their masses and mixing angles. One major difficulty in implementing this goal is the uncertain origin of supersymmetry breaking which is important in triggering various stages of gauge symmetry breaking, including the electroweak breaking stage. As I shall discuss, however, with a plausible model for this supersymmetry breaking quite realistic mass matrices may emerge in specific models.

382 G. G. ROSS

5. Models for low-energy physics

The ideal scenario in choosing a model for low-energy physics is to examine the possible vacuum configurations of the string and ask which one will be chosen in the evolution of the universe. Some progress has been made in this direction, but certainly not sufficient to choose between the many candidate vacua. It therefore seems reasonable, at this stage, to invoke phenomenological criteria when selecting a vacuum solution. The most obvious constraint is that this solution should reproduce the correct multiplet structure at low energies, and this leads to the consideration of three-generation models. It should also be supersymmetric below the compactification scale if we are to have a hope of solving the hierarchy problem. Processes violating baryon-number and lepton-number in such models can proceed very quickly and there should be a mechanism for suppressing these processes.

In this section I discuss four promising three-generation schemes to illustrate just how far one can get towards a viable theory from a definite vacuum solution of the string equations of motion.

5.1. Calabi–Yau models

Of the known Calabi–Yau models, only one distinct variety (of the CICY type) is known which has just three generations (Aspinwall *et al.* 1987; Candelas *et al.* 1988; Green *et al.* 1987 *a*). Three different manifold constructions of the model are known based on $CP_2 \times CP_2$, $CP_3 \times CP_3$ (Yau 1985) and $CP_3 \times CP_2$ (Schimmrigk 1987), respectively. The simplest, which does not require any blowing-up of singularities, starts with $R_0 = CP_3 \times CP_3$. We call its complex coordinates $x_i, y_i, i = 0, 1, 2, 3$. The three-generation model is based on the quotient manifold $R = R_0/G$, where G is the freely acting Z_3 discrete group defined by

$$\left. \begin{aligned} g &: (x_0\, x_1\, x_2\, x_3) \to (x_0\, \alpha^2 x_1, \alpha x_2, \alpha x_3), \\ g &: (y_0\, y_1\, y_2\, y_3) \to (y_0, \alpha y_1, \alpha^2 y_2, \alpha^2 y_3). \end{aligned} \right\} \tag{5.1}$$

The six complex coordinates are reduced to the three needed for a space of $SU(3)$ holonomy by the constraints,

$$\left. \begin{aligned} &\sum x_2^3 + a_1 x_0 x_1 x_2 + a_2 x_0 x_1 x_3 = 0, \\ &\sum y_2^3 + b_1 y_0 y_1 y_2 + b_2 y_0 y_1 y_3 = 0, \\ &x_0 y_0 + c_1 x_1 y_1 + c_2 x_2 y_2 + c_3 x_3 y_3 + c_4 x_2 y_3 + c_5 x_3 y_2 = 0. \end{aligned} \right\} \tag{5.2}$$

The model has vanishing first Chern class, an Euler characteristic of -6 corresponding to three generations and Hodge numbers (Greene *et al.* 1986, $h_{2,1} = 9$ and $h_{1,1} = 6$ corresponding to nine copies of 27 and six of $\overline{27}$. The model has a complex structure determined by the $h_{2,1} = 9$ coefficients of equation (4.2) that should be interpreted as VEVs of the scalar components of the nine chiral supermultiplets the $h_{2,1}$ moduli (Ross 1988 *b*). For a special choice of these parameters $a_i = b_i = 0$ and only c_{ii} non-vanishing with $c_{00} = c_{11} = 1$, $c_{22} = c_{33} = c$ the manifold has a large set of discrete symmetries (Greene *et al.* 1986) that imply the existence of several nearly flat directions among the gauge non-singlet fields. It can be shown that after supersymmetry breaking this choice corresponds to a local minimum of the potential describing the moduli (Ross 1988 *b*) showing how a particular complex structure may be chosen.

The phenomenology of this model (Greene *et al.* 1986, 1987) has been examined assuming this symmetric point and also assuming a non-trivial embedding of the Z_3 equation (5.1) in the gauge group. (There are only three possible such embeddings so this assumption does not amount to 'fine-tuning' of the vacuum structure.) With this embedding the gauge group is broken to $[SU(3)]^3$ and the multiplet structure consists of nine generations plus six antigenerations of leptons in the $(1, 3, \bar{3})$ representation plus seven generations and four antigenerations of quarks in both the $(3, \bar{3}, 1)$ and $(\bar{3}, 1, 3)$ representations. The flat directions allow for just two stages of large gauge symmetry breaking (i.e. very much greater than M_W) reducing the gauge group to $SU(3) \times SU(2) \times U(1)$. This is because after these two stages all $\overline{27}$ components have acquired large masses so there are no further D flat directions along which large vevs may develop. Because the trilinear couplings $(\overline{27})^3$ are completely known (Distler et al. 1987) in this model it is possible to compute the radiative corrections and verify that supersymmetry breaking effects will trigger the gauge symmetry breaking along the flat directions (Yamagishi 1983; Alvarez-Gaume et al. 1983; Ibañez et al. 1985; Gato et al. 1985). The states left light down to the electroweak breaking scale consist of just three generations of quarks and leptons, together with their superpartners, one pair of Higgs doublets (H_1^0, H_1^-), (H_2^+, H_2^0), the minimum necessary to generate electroweak breaking, together with their superpartners. The only additional states are those in three neutral and one charged lepton superfield whose masses are of order $1\,\mathrm{TeV}$. The new $Q = -\frac{1}{3}$ colour states in triplet the $(3, \bar{3}, 1)$ and $(\bar{3}, 1, 3)$ representations all acquire the large gauge breaking mass.

5.2. Orbifold models

Much work has gone into the construction of orbifold models both with $(2, 0)$ and $(2, 2)$ world-sheet supersymmetry and realistic versions have now been developed. These models rely on flat directions to reduce the rank of the gauge group at a high scale. A typical version of such models (Ibañez et al. 1987 a–c, 1988) starts with a Z_3 orbifold with non-trivial embedding of the point group in the gauge group leading to a gauge group after compactification,

$$SU(3) \times SU(2) \times U(1)^8 \times SO(10), \tag{5.3}$$

and a spectrum consisting of three generations plus numerous exotic particles. The symmetries are found to require the existence of numerous flat directions along which the vevs of various fields in the theory are undetermined. A selection criterion is imposed by hand that requires that vevs should only develop along flat directions which leave the $U(1)_Y$ of the standard model unbroken. If large vevs along these allowed directions are assumed, then it is found that the unwanted colour triplets in the model, which may mediate proton decay, became massive along with many others of the extra particles. In this respect the model is similar to the Calabi–Yau model discussed above.

The origin of the large vevs in these models follows from the fact that they always have one anomalous $U(1)_X$, which gives a non-vanishing contribution $(g/192\pi^2)\,\mathrm{Tr}\,[X]$ to the associated D term. Then the complete D term is given by

$$D^X = \Sigma_a\, q_a^X |\Sigma_a|^2 + \frac{g}{192\pi^2}\,\mathrm{Tr}\,[X], \tag{5.4}$$

where Σ_a are the scalar components of chiral superfields in the theory and q_a^X is their X charge. Minimization of the vacuum energy will require that some of the fields Σ_a develop vevs of the order of the compactification scale, the minimum of the potential corresponding to vevs developing along the 'flat' directions. Note, however, that there are further flat directions in general whose vevs are not determined by the anomalous F term, as they happen to be invariant under X.

To proceed further, it is necessary to select one among the many candidate flat directions, assigning arbitrary vevs. When they are not determined by the condition, D^X vanishes. Along a phenomenologically promising choice chosen by 'hand' with gauge group $SU(3) \times SU(2) \times U(1)_Y$, it is found that all exotic quarks and leptons get a large mass, leaving just the quark and lepton states needed for the supersymmetric standard model plus three pairs of doublet and antidoublet Higgses and some massless hypercharge singlets. In addition, there is an $SO(10) \times U(1)$ hidden sector. Again, the resultant structure is very similar to the Calabi–Yau model discussed above.

The phenomenological implications of the models have been broadly studied (Ibañez *et al.* 1987*a–c*, 1988; Bailin *et al.* 1987; Casas *et al.* 1987; Casas & Muñoz 1988). It is found that proton decay is inhibited because of the absence of $q\bar{q}l$ couplings (see §6.1). There may be an additional light $U(1)'$ gauge boson coupled only to leptons, its mass is determined by vevs not fixed by equation (5.4). If this is present, it guarantees the absence of the $q\bar{q}l$ couplings; if not, the absence of $q\bar{q}l$ follows only if the particular flat direction considered is chosen. Quark and lepton masses arise after electroweak breaking provided the relevant Yukawa couplings to the Higgs doublets are present. At tree level, the symmetries of the model allow only one set of Yukawa couplings, so that only the heaviest quarks and leptons (t, b, c, τ) can acquire tree-level masses. It may be that the remaining states acquire masses through loop corrections, although this has not been studied and it is difficult to see how sufficiently large masses will come from such a mechanism.

It is interesting that, again, a multiplet structure very close to the supersymmetric standard model is obtained in a specific model, but once more the outstanding question is why the particular flat direction is chosen. The generic expectation seems to be that the large rank of the gauge group normally found in orbifolds will be reduced due to the flat directions, although it is of course, in principle, possible that additional light $U(1)$ gauge bosons persist in low-energy scales.

5.3. *Flipped* $SU(5)$ *(Antoniadis et al.* 1987*b)*

Early attempts to construct viable models based on the heterotic superstring used non-trivial gauge group embedding of the quotient group to reduce the gauge group from $E_8 \times E_6$. It was quickly realized that the maximal size of the resultant gauge group is severely constrained for, in general, the subsequent breaking at a high scale of the gauge symmetry by the usual Higgs mechanism will leave an $SU(5)$ group intact and the $SU(5)$ gauge interactions then generate nucleon decay at an unacceptable rate. The reason for this is obvious with the conventional E_6 assignment of particles, for the only states which are singlets under the standard model gauge group, $SU(3) \times SU(2) \times U(1)$ (these are the only states which can receive large (very much greater than M_W) vevs) are ν_R and N, and these leave an $SU(5)$ subgroup of E_6 intact. The trouble is that the breaking of $SU(5)$ in this case requires an adjoint representation of Higgs scalars, but in superstring models the chiral supermultiplets do not come in adjoint representations.

There is an exception to this rule, the so-called 'flipped' $SU(5) \times U(1)$, in which the assignments of the u_L and d_L quarks to $SU(5)$ representations are interchanged (Barr 1982; Derendinger *et al.* 1984). This is not possible in usual $SU(5)$ because the charge operator is a generator of SU(5), so the charge should be traceless over a representation, but in SU(5) $\times$ U(1) there is a U(1) component of the charge allowing for this change. Moreover, in the breaking of E_6 it is not unreasonable that such additional U(1) factors should appear. With this assignment, it is possible to break SU(5) at a high scale down to the standard model group by using just 5, $\bar{5}$, 1 and 10 representations of SU(5), and these representations may be present in superstring compactification.

There is a potential problem for such a grand unified scheme that did not arise in the problems so far discussed, namely the doublet–triplet splitting problem. This arises because a grand unified group such as SU(5) requires the existence of Higgs colour triplets as degenerate partners of the Higgs doublets. These, and their fermion superpartners, can mediate proton decay and should be made very massive, of the order of 10^{16} GeV, while keeping their doublet partners, needed for electroweak breaking, light, of $O(M_\mathrm{W})$. The superstring theories so far discussed neatly avoid this problem, for their initial gauge groups are *smaller* than SU(5) and so imply no symmetry relation between SU(2) doublets and SU(3) triplets. This represents a significant improvement over the general grand unified theory (GUT).

In the case of a flipped SU(5), however, the problem is present, but happily is solved by the missing partner mechanism (Masiero *et al.* 1982; Grinstein 1982; Kounnas *et al.* 1983; Campbell *et al.* 1987), in which the couplings of the 5 and $\bar{5}$ of Higgses contain just the required triplet mass term but no doublet mass term. This is not in conflict with the underlying SU(5) symmetry, for the symmetry is broken by the large vev of the v_R component of the 10, and only the colour triplets couple to the v_R component.

Flipped SU(5) has been discussed in the context of manifold compactification (Antoniadis *et al.* 1988 *a*), but no explicit realization is known. Recently, a four-dimensional string construction with flipped SU(5) has been developed (Antoniadis *et al.* 1988 *b*) that predicts the multiplet structure as well as the gauge group. The most promising version has an 'observable' GUT group SU(5) $\times$ U(1) $\times$ U(1)^3 with three generations of matter field with SU(5) $\times$ U(1) transform properties:

$$F_i = (10/\tfrac{1}{2}), \quad \bar{f}_i = (5, -\tfrac{3}{2}), \quad T_1^\mathrm{c} = (1, \tfrac{5}{2}); \quad i = 1, 2, 3.$$

In addition there are the 10-dimensional Higgses needed to break SU(5) $\times$ U(1) down to the standard group,
$$H_{1,2} = (10, \tfrac{1}{2}), \quad \bar{H}_{1,2} = (10, -\tfrac{1}{2}),$$

and the five-dimensional Higgses needed to break the standard model,
$$h_{1,2,3} = (5, -1), \quad \bar{h}_{1,2,3} = (\bar{5}, 1).$$

There are also three pairs of singlets and their complex conjugates:
$$\varphi_{12} = (1, 0), \quad \varphi_{23} = (1, 0), \quad \varphi_{13} = (1, 0).$$

The model requires a stage of large symmetry breaking, which occurs along a flat direction. Unlike the previous models there are not a large number of flat directions to consider, so it is easier to establish the breaking pattern. There is, in addition to the visible sector multiplets, an '*à la carte*' hidden sector that may be responsible for supersymmetry breaking, although its properties and flat directions have not been explored to date.

After GUT breaking, one is left with basically the supersymmetric standard model, possibly with the addition of extra U(1) gauge bosons and Higgs doublets and singlets; the details have not been completely determined. The GUT-breaking scale is large, $M_G \approx 10^{16 \pm 1}$ GeV, the dominant nucleon decay is via dimension-six operators (see §6.1) and is expected to be very slow $(10^{33}y \ll 10^{57}y)$.

Fermion masses are determined by undefined Yukawa couplings, but some general properties of the masses are known (Antoniadis *et al.* 1988*b*; Leontaris & Nanopoulos 1988; Leontaris 1988). There is an SO(10) relation $m_b = m_t$ at the Planck scale which, after QCD renormalization, gives m_b of $O(6\ \mathrm{GeV})$. The top quark mass is found to be of $O(70\ \mathrm{GeV})$ if electroweak breaking via radiative corrections is to be triggered. The structure of the superpotential consistent with the known symmetries of the model can give an acceptable pattern of fermion masses.

5.4. *Construction by using $N = 2$ superconformal theories*

Recently Gepner (Gepner 1987) has constructed a version of the three-generation Calabi–Yau theory by using the tensor product of one $k = 1$ theory and three $k = 16$ theories. The gauge group before gauge symmetry breaking is $E_8 \times E_6 \times U(1)^3$ and the multiplet structure is in agreement with that found in the Calabi–Yau version. In addition, he is able to determine that there are 62 gauge singlet chiral superfields, three of which are needed to give mass to the three additional gauge bosons as one moves away from the exceptional point in moduli space. The symmetries of the model do not, however, correspond to the model constructed above, but rather to the model constructed by Schimmenck (1987) using the $CP_2 \times CP_3$ manifold.

The advantage of this version of the model is that all the trilinear couplings have been calculated, and extensive information about the E_6 gauge-singlet structure and the symmetries has been derived (Sotkou & Stanishkov 1988; Greene *et al.* 1988).

In constructing this version of the model it is necessary to mod out by a $Z_3 \times Z_3$ group, the first Z_3 being freely acting and the second not (in the manifold construction the resultant singularities must be resolved). If I also assume a non-trivial embedding of the first Z_3 in the gauge group we get a gauge group $SU(3)^3 \times U(1)^3$ and the same multiplet structure given as I discussed in §5.1. However, if the second Z_3 is embedded in the gauge group, although the same gauge group obtains, a different multiplet structure results for the quarks for only three generations survive in the $(3, \bar{3}, 1)$ and $(\bar{3}, 1, 3)$ representations. So far only this latter case has been analysed (Lutken & Ross 1989) using the new information on symmetries and couplings coming from the conformal field theory construction. In this case it is straightforward to determine whether there are any stages of symmetry breaking subsequent to compactifications by studying the renormalization group equations. Again it is found there will be gauge breaking and there is a local minimum in which the gauge group is reduced to $SU(3) \times SU(2) \times U(1)$.

6. PHENOMENOLOGY OF THE LOW-ENERGY MODEL

Having discussed the expectations for the low-energy multiplet structure in general and in particular, I turn now to a résumé of the low-energy phenomena to be expected from string models in four flat space-time dimensions.

6.1. *Baryon-number and lepton-number violation*

An essential ingredient for a viable low-energy model is that violation of the baryon and lepton numbers should be within experimental limits. However, all supersymmetric models suffer from possible baryon-number violation through dimension-four operators of the form

$$[lh_1, lle^c, qld^c, u^c d^c d^c]_F, \tag{6.1}$$

where the superfields l and q refer to left-handed (LH) doublets of leptons and quarks under $SU(2)_L$; e^c, u^c and d^c refer to LH singlet antileptons and antiquarks, respectively; and h_1, h_2 are LH Higgs superfields. The presence of these operators leads to processes violating lepton and baryon number unsuppressed by an inverse powers of mass greater than the supersymmetry-breaking scale ($O(1\text{ TeV})$), and hence some, or all, of these operators should be absent at tree level (Weinberg 1982; Sakai & Yanagida 1982). To achieve that it is normally assumed that there is a discrete symmetry of the superpotential, called matter parity, under which quark and lepton superfields change sign, while all other superfields remain the same. This forbids operators of the type given in equation (5.1), while allowing the operators needed for mass generation, namely,

$$[lh_2 e^c, qh_2 d^c, qh_1 u^c, h_1 h_2]_F, \tag{6.2}$$

where h_1 is the LH Higgs supermultiplet of the standard model needed to give u-quarks a mass, and h_2 is the one needed for d-quarks and leptons. The last term in (6.2) is needed to allow for an acceptable pattern of vEVs for h_1 and h_2.

Although original expectations were that it would be difficult to find string models that achieved the necessary suppression that has not proved to be the case because of the additional gauge and discrete symmetries predicted by the compactification. In the Calabi–Yau model (Greene *et al.* 1986, 1987) and also in the $N = 2$ superconformal version (Lutken & Ross 1989) R parity emerges naturally. For the Calabi–Yau model of §5.1 there is a Z_2 group with the action

$$x_2 \leftrightarrow x_3; \quad y_2 \leftrightarrow y_3. \tag{6.3}$$

This clearly leaves the defining equation (5.2) invariant for the maximally symmetric manifold. For the $N = 2$ theory of §5.4 there is a Z_2 group under which twisted modes are odd and untwisted even. Although both of these symmetries are spontaneously broken after compactification they combine with a Z_2 subgroup of the residual gauge group to leave an unbroken R parity as required. In the orbifold and fermionic string model it turns out that gauge-symmetries prevent the appearance of the operators of (5.1) (Ibañez *et al.* 1987*a–c*, 1988; Bailin *et al.* 1987; Casas *et al.* 1987; Casas & Muñoz 1988; Antoniadis *et al.* 1988*b*).

Although the models have eliminated dimension-four baryon-number violation, at dimension five and beyond there *are* baryon- and lepton-number violating terms involving light states.

Let us first consider baryon-number violation mediated by the D-quarks of the model. The dimension-five operator contribution gives a proton decay amplitude proportional to M_D^{-1}. The calculation of the proton decay rate following from this contribution involves considerable uncertainty due to the unknown Yukawa couplings and supersymmetry-breaking effects, but it has been plausibly argued (Weinberg 1982; Greene *et al.* 1988) that these graphs will lead to unacceptable rates for proton decay unless the massive colour triplet states mediating the decays satisfy the bounds $M_D > 10^{16}$ GeV. Studies of the potential governing the large breaking scale suggest that breaking scales large enough to satisfy this bound are quite reasonable.

In the flipped SU(5) model the dimension-five terms are absent so proton decay proceeds only via dimension-six operators with amplitude proportional to M_G^2 and likely to be invisible.

Lepton-number-violating processes for the models turn out to be well within current experimental limits for breaking scales consistent with baryon-number-violation limits.

6.2. *Neutrino masses*

The neutrinos belong to the lepton doublets and can acquire Dirac masses by coupling with the singlet-lepton components. These mass terms arise only after electroweak breaking from terms allowed by the discrete symmetries, which may be expected to be of the same order as the charged-lepton masses m_l. In addition, we must consider the effect of the large Majorana mass terms for the singlet-lepton components. The usual diagonalization of the neutrino mass matrix leaves light neutrinos with Majorana mass of the order of m_l^2/m^*, where m^* is the mass of the singlet component with Dirac coupling to the neutrino.

In models with a large-scale, M_I, of gauge symmetry breaking the expectation is that Majorana mass will be large, coming via dimension-four non-renormalizable terms in the superpotential and giving m^* of the order of M_I^2/M_c. This gives (Greene *et al.* 1986, 1987; Nandi & Sarkar 1986) neutrino masses of the order of $m_t^2 m_c/M_I^2$. For $m_1 = 10^{16}$ GeV, this is of order 10^{-11}, 10^{-7} and 10^{-5} eV for the e, μ or τ, respectively, well within current bounds on Majorana neutrino masses.

6.3. *The weak mixing angle*

After the compactification the gauge couplings remain related as if embedded in a GUT. This means that the weak mixing angle is fixed, and for E_6-based models the prediction is $\sin^2\theta_w = \frac{3}{8}$. However, this value gets renormalized as one continues the prediction from the compaction scale to the low-energy scale as a result of the effect of the light states in the theory. These are readily calculated. For example in the Calabi–Yau model (Greene *et al.* 1986) with a gauge breaking scale of not less than $O(10^{16}$ GeV) consistent with the constraint of baryon-number violation $\sin^2\theta_w = 0.225$, in good agreement with the experimental values. Reasonable values for $\sin^2\theta_w$ come too in various orbifold models and the flipped SU(5) model discussed above (Ibañez *et al.* 1987 *a–c*, 1988; Bailin *et al.* 1987; Casas *et al.* 1987; Casas & Muñoz 1988; Antoniadis *et al.* 1988 *b*).

6.4. *CP violation*

Dine & Seiberg (1986) have pointed out that strong CP violation is a generic problem for superstring models owing to small instanton contributions being enhanced by the growth of the strong coupling near the compactification scale. To avoid this, one may a try to use a Peccei–Quinn (PQ) symmetry allowing for the cancellation of the CP violation by the relaxation of the associated axion. However, Calabi–Yau compactification has no continuous (non-gauged) symmetries, so apparently there is no possibility of a PQ symmetry and hence of suppressing strong CP violation in such schemes (Choi & Kim 1985). However, the discrete symmetries of the model may give rise to an approximate PQ symmetry (broken only by non-renormalizable terms), which is quite adequate for suppressing CP violation (Lazarides *et al.* 1986; Casas & Ross 1987). Moreover, the natural size of PQ breaking is of $O(10^{11}$ GeV), which gives an easy explanation of the scale needed for a viable invisible axion. It remains to be seen whether this mechanism is realized in the three-generation models discussed here.

6.5. *Light Higgs scalars and electroweak breaking*

It is no surprise that three families of quarks and leptons remain light, because until electroweak breaking, the states in unpaired 27s (three families by construction) must remain massless. However, the Higgs doublets $H_1 = (h_1^+, h_1^0)$ and $H_2 = (h_2^0, h_2^-)$ needed for electroweak breaking; may couple in an $SU(3) \times SU(2) \times U(1)$ invariant way, giving H_1 and H_2 mass, and so it is surprising that any such components of a 27 remain light after large-scale breaking. The reason one pair escapes a large mass is that the symmetries of the model prevent an H_1–H_2 interaction until a high (non-renormalizable) order.

In the Calabi–Yau models the residual discrete symmetries guarantee this (Greene *et al.* 1986), whereas in the flipped $SU(5)$ model, it is the missing partner mechanism (Masiero *et al.* 1982; Grinstein 1982; Kounnas *et al.* 1983) following from the gauge symmetries that keeps the Higgs light. In the orbifold models, too, it is the residual gauge symmetry that protects the light Higgs.

The triggering of electroweak breaking proceeds in the usual way via supersymmetry breaking from the hidden sector generating a negative squared mass, $-m^2$, for the Higgs scalars. This time, however, there is no D flat direction as there are no residual $\overline{27}$ components left light and so the potential $V(H)$ has the form $-m^2|H|^2 + g^2|H|^4$ leading to the expectation that the vev of H should be of $O(m)$, the supersymmetry breaking mass scale. This scale is related to the scale at which a gaugino condensate occurs in the hidden sector, i.e. where the hidden-sector gauge coupling becomes large. This in turn is related to the compactification scale times the factor $\exp(-3S/b_0)$, where b_0 is the coefficient of the hidden sector β function and S is a field related to the dilation field and the field setting the scale of compactification, which in turn sets the scale for the gauge coupling. For large S/b_0 a large hierarchy of masses may arise and model calculations suggest (Ross 1988) it is possible to achieve the very small electroweak breaking scale required.

6.6. *Quark and lepton masses*

Perhaps the most exciting prospect for compactified string theories is that they should determine the quark and lepton mass matrices allowing for a precise test of the theory. In practice this is often difficult to implement for the techniques needed to determine the couplings have not been developed in every case. Within these limitations the mass spectra in the various models have been explored with encouraging results. As discussed in §5 the orbifold (Ibañez *et al.* 1987a–c, 1988; Bailin *et al.* 1987; Casas *et al.* 1987; Casas & Muñoz 1988), and fermionic models (Antoniadis *et al.* 1988b; Leontaris & Nanopoulos 1988; Leontaris 1988) may give rise to reasonable mass matrices, though no definite predictions have been determined. In the Calabi–Yau model, too, a reasonable mass spectrum emerges (Greene *et al.* 1986, 1987) and even a satisfactory prediction relating mixing angles results. However, the only model whose predictions are well established is the tensor product model of §5.4, so we will concentrate on this model to illustrate how the mass matrices are determined (Lutken & Ross 1989).

The first step is to determine the light particle spectrum after the stages of large spontaneous-gauge symmetry breaking. This consists of just three quark doublets, the extra D quarks getting large mass via the known trilinear couplings. In addition there is left light one pair of Higgs doublets H_1 and H_2 given by

$$\left.\begin{aligned}
{[H_1]}_i &= [a_3\lambda_3 + a_5\lambda_5 + a_1\lambda_1]_{(i,1)}, \\
{[H_2]}_i &= [a_3'\lambda_3 + a_5'\lambda_5 + a_1'\lambda_1 + a_2'\lambda_2]_{(i,2)},
\end{aligned}\right\} \tag{6.4}$$

[71]

where λ_1, λ_3 and λ_5 are three of the original nine $(1, 3, \bar{3})$ multiplets and are mixing angles determined by the Yukawa couplings and the large vevs. The reason the primed and unprimed values may differ is due to the fact that with two stages of spontaneous symmetry breaking along $(\lambda)_{(1, 3, 3)}$ and $(\lambda)_{(1, 3, 2)}$ directions there may be masses $\lambda_{(i, 1)} \lambda_{(i, 2)} \langle \lambda_{(3, 3)} \rangle$ and $\lambda_{(i, 1)} \lambda_{(i, 3)} \langle \lambda_{(3, 2)} \rangle$ destroying the symmetry between the components of (6.4). This has the important consequence that up and down quark mass matrices may differ.

Finally, there are left light three lepton doublets, mixtures of the nine $(1, 3, \bar{3})$ multiplets and their neutrino singlet partners together with three heavy leptons.

Following from equation (6.4) and the determination of the Yukawa couplings, k_i, it is easy to write down the quark mass matrices. For the up quarks,

$$[Q_1 \, Q_2 \, Q_3] \begin{bmatrix} 0 & a_5 k_0 & 0 \\ 0 & 0 & a_1 k_0 \\ a_5 k_1 & a_1 k_0 & a_3 k_2 \end{bmatrix} \begin{bmatrix} q_1 \\ q_2 \\ q_3 \end{bmatrix}. \tag{6.5}$$

Here $k_0 = 0.6$, $k_1 = 0.5$, $k_2 = 1.05$ are the non-vanishing Yukawa couplings. If I fix the constants a_i by the three up quark masses we obtain predictions for the mixing angles of the (left-handed) up quarks. Doing the same for the down quarks gives finally a mixing matrix of the form.

$$\begin{bmatrix} 1 & \lambda & A\lambda^2\rho \\ -\lambda & 1 & A\lambda^2 \\ A\lambda^3(1-\rho) & -A\lambda^2 & 1 \end{bmatrix}, \tag{6.6}$$

where
$$\left. \begin{aligned} A\lambda^2 &= m_{\rm d}/m_{\rm s} - m_{\rm u}/m_{\rm c} = 0.04 \pm 0.002, \\ A\lambda\rho^2 &= m_{\rm d}/m_{\rm b} - m_{\rm u}/m_{\rm t} = 0.8 \pm 0.4, \end{aligned} \right\} \tag{6.7}$$

the numbers being experimental measurements. For $m_{\rm b} = 5$ Gev, $m_{\rm s} = 150$ MeV, $m_{\rm d} = 7.5$ MeV we get excellent agreement.

The magnitude of a_i, a_i' and hence the quark masses are related to the large vevs breaking the gauge symmetry. These are not entirely determined by the calculated trilinear terms but depend also on the non-renormalizable terms $(27 \cdot \overline{27})^n$. However, the vevs have upper bounds corresponding to the scales at which the m^2 becomes negative. These scales are, in turn, determined by the trilinear couplings leading to the result, if we assume the upper bounds are saturated;

$$a_i/a_{j\alpha} \sim \exp\left\{ -4\pi^2 [1/\sum_\alpha c_\alpha (k_i^\alpha)_\beta^2 - 1/\sum c_\beta (k_j^\beta)^2] \right\}, \tag{6.8}$$

where k_i^α is the Yukawa coupling involved in the one-loop corrections with virtual states α driving the evolution of m^2 and c_α are constants associated with the Feynman graph. The quark masses are given in terms of the a_i: $m_{\rm c}/m_{\rm t} = a_5 k_0/a_3 k_2$, $m_{\rm u}/m_{\rm c} = a_1 a_k k_0^2/a_3 k_2$ and so it is easy to get large hierarchies of masses from the exponential factors in (6.8). Detailed predictions require a more careful treatment of the renormalization group equations than given in (6.8) and this has not yet been done but the expectation is that $m_{\rm t}$ will be of $O(M_{\rm w})$.

7. SUMMARY AND CONCLUSIONS

The attempts to determine the low-energy model from the superstring are beset by the problems involved in determining the several stages of symmetry breaking, both of the underlying space-time, and of the supersymmetry and gauge symmetries. As a result of these problems, the candidate four-dimensional low-energy theories are very numerous. What is needed is an understanding of the mechanics, choosing a (unique?) vacuum state. In this, supersymmetry-breaking effects must play a crucial role when a selection is made among the many degenerate supersymmetric vacuum solutions.

Unfortunately, our understanding of supersymmetry breaking is limited, both because it requires a non-perturbative trigger and because the full form of the effective lagrangian, including string effects and heavy Kaluza–Klein exchange effects, is not known. However, some progress has been made; for example, it is known how supersymmetry-breaking effects can select a particular complex structure, thus choosing one of a class of Calabi–Yau compactifications.

In the absence of a complete understanding of the choice of vacuum state, it seems reasonable to use phenomenological criteria to select a promising scheme, and to investigate that scheme in detail as an example of the problems and answers that may be expected from the 'true' compactification.

The low-energy phenomenology of the candidate three-generation models analysed have certain features in common. All appear to have a large scale of symmetry breaking after compactification, leading to the possibilities of structure in the mass matrices. It also appears likely that this breaking will substantially reduce the low-energy spectrum; for example, it is no longer thought so likely that there will be additional $U(1)$ gauge bosons at low energies, beyond the standard model $SU(3) \times SU(2) \times U(1)$ structure. Similarly, the presence of new exotic fermion states is no longer mandated by string compactification.

The detailed structure of the mass matrices in the tensor product model analysed has the correct form to explain the mixing angle and the hierarchy of fermion masses although it remains to be seen whether a complete analysis of the renormalization-group equations will give, in detail, good post-dictions for these quantities. At the very least the analysis shows how superstring models may realize their promise and give predictions going beyond the standard model.

It would be ridiculous, at this stage, to claim that any one model is *the* effective low-energy model following from the superstring, but its existence does show that the low-energy structure emerging from an underlying compactified theory can closely resemble the observed world. Moreover, it shows how the underlying structure may realize its promise and provide answers to the questions posed by the standard model, namely the origin of its multiplet structure and its parameters, the gauge and Yukawa couplings, the masses, and the mixing angles.

References

Alvarez-Gaume, L., Polchinski, J. & Wise, M. 1983 *Nucl. Phys.* B **221**, 495.

Antoniadis, I., Bachas, C. & Kounnas, C. 1987*a Nucl. Phys.* B **289**, 87.

Antoniadis, I., Ellis, J., Hagelin, J. S. & Nanopoulos, D. V. 1987*b Phys. Lett.* B **194**, 231.

Antoniadis, I., Ellis, J., Hagelin, J S & Nanopoulos, D, V, 1988*a Phys. Lett.* B **205**, 459.

Antoniadis, I., Ellis, J., Hagelin, J. S. & Nanopoulos, D. V. 1988*b* Madison–CERN preprint MAD/TH/88-10 and CEN-TH.5005/88.

Aspinwall, P. S., Greene, B. R., Kirklin, K. H. & Miron, P. J. 1987 University of Oxford preprint 26/87.

Bailin, D., Love, A. & Thomas, S. 1987 *Nucl. Phys.* B **288**, 431; *Phys. Lett.* B **194**, 385.

Barr, S. M. 1982 *Phys. Lett.* B **112**, 219.

Binétruy, P. & Gaillard, M. K. 1986 *Phys. Lett.* B **168**, 347.

Breit, J. D., Ovrut, B. A. & Segrè, G. C. 1985*a Phys. Lett.* B **158**, 33.

Breit, J. D., Ovrut, B. A. & Segrè, G. C. 1985b *Phys. Lett.* B **162**, 30.
Burgess, C. P., Font, A. & Quevedo, F. 1986 *Nucl. Phys.* B **272**, 661.
Campbell, B., Ellis, J., Hagelin, J. S., Nanopoulos, D. V. & Ticciati, R. 1987 *Phys. Lett.* B **198**, 200.
Candelas, P. 1987 University of Texas at Austin preprint UTTG-05-87.
Candelas, P., Dale, A. M., Lütken, C. A. & Schimmrigk, R. 1988 *Nucl. Phys.* B **298**, 493.
Candelas, P., Horowitz, G., Strominger, A. & Witten, E. 1985 *Nucl. Phys.* B **258**, 46.
Casas, J. A., Katechou, E. K. & Muñoz, C. 1987 University of Oxford preprint 1/88.
Casas, J. A. & Muñoz, C. 1988 University of Oxford preprints 28/88 and 32/88.
Casas, J. A. & Ross, G. G. 1987 *Phys. Lett.* **192**, 119.
Chamseddine, A. & Derendinger, J.-P. 1987 Preprint ETH-PT/87-2.
Choi, K. & Kim, J. E. 1985 *Phys. Lett.* B **154**, 393.
Coughlan, G. D., German, G., Ross, G. G. & Segrè, G. G. 1988 *Phys. Lett.* **198**, 467.
Cremmer, E., Ferrara, S., Kounnas, C. & Nanopoulos, D. V. 1983 *Phys. Lett.* B **133**, 61.
Cvetic, M. 1987 SLAC preprint, SLAC-PUB-4325.
del Aguila, F., Blair, G., Daniel, M. & Ross, G. G. 1986 *Nucl. Phys.* B **272**, 413.
Derendinger, J.-P., Kim, J. E. & Nanopoulos, D. V. 1984 *Phys. Lett.* B **139**, 170.
Derendinger, J.-P., Ibañez, L. E. & Nilles, H. P. 1985 *Phys. Lett.* B **155**, 65.
Dine, M. & Seiberg, N. 1986 *Nucl. Phys.* B **273**, 109.
Dine, M., Kaplunovsky, V., Mangano, M., Nappi, C. & Seiberg, N. 1985a *Nucl. Phys.* B **259**, 519.
Dine, M., Rohm, R., Seiberg, N. & Witten, E. 1985b *Phys. Lett.* B **156**, 55.
Distler, J., Greene, B., Kirklin, K. & Miron, P. 1987 University of Oxford preprint 37/87.
Distler, J. & Greene, B. 1988 Cornell preprint CLNS 88/834, Harvard preprint HUTP-88/A020.
Dixon, L. 1987 Princeton preprint PUPT-1074. (To be published in *Proc. 1987 ICTP workshop in High Energy Physics and Cosmology*.)
Dixon, L., Harvey, J., Vafa, C. & Witten, E. 1985 *Nucl. Phys.* B **261**, 620; B **274**, 285.
Dixon, L., Harvey, J., Vafa, C. & Witten, E. 1986 *Nucl. Phys.* B **274**, 285.
Ellis, J., Kounnas, C. & Nanopoulos, D. V. 1984a *Nucl. Phys.* B **241**, 406; B **247**, 373.
Ellis, J., Lahanas, A. B., Nanopoulos, D. V. & Tambakis, K. 1984b *Phys. Lett.* B **134**, 429.
Ferrara, S. L., Girardello, X. X. & Nilles, H. P. 1983 *Phys. Lett.* B **125**, 457.
Gato, B., Leon, J., Perez-Mercader, J. & Quiros, M. 1985 *Nucl. Phys.* B **253**, 285.
Gepner, D. 1987 Princeton University preprint.
Gepner, D. 1988 *Nucl. Phys.* B **296**, 757.
Gildener, E. 1976 *Phys. Rev.* D **14**, 1667
Gildener, E. 1980 *Phys. Lett.* B **92**, 111.
Gildener, E. & Weinberg, S. 1976 *Phys. Rev.* D **15**, 3333.
Green, M. B., Schwarz, J. H. & Witten, E. 1987b *Superstring theory*. Cambridge University Press.
Green, P., Hübsch, T. & Lütken, C. A. 1987a All Hodge numbers of all complete intersection Calabi–Yau manifolds, CERN-TH. 4933/87.
Greene, B. R., Kirklin, K. H., Miron, P. J. & Ross, G. G. 1986 *Nucl. Phys.* B **278**, 667; *Phys. Lett.* B **180**, 69.
Greene, B. R., Kirklin, K. H., Miron, P. J. & Ross, G. G. 1987 *Nucl. Phys.* B **292**, 606.
Greene, B. R., Lutken, A. & Ross, G. G. 1988 Nordita preprint.
Grinstein, B. 1982 *Nucl. Phys.* B **206**, 387.
Hübsch, T. 1987 *Communs math. Phys.* **108**, 291.
Ibañez, L. E., Lopez, C. & Muñoz, C. 1985 *Nucl. Phys.* B **256**, 218.
Ibañez, L. E., Kim, J. E., Nilles, H. P. & Quevedo, F. 1987a *Phys. Lett.* B **191**, 3.
Ibañez, L. E., Nilles, H. P. & Quevedo, F. 1987b *Phys. Lett.* B **187**, 25.
Ibañez, L. E., Nilles, H. P. & Quevedo, F. 1987c *Phys. Lett.* B **192**, 332.
Ibañez, L. E., Mas, J., Nilles, H. P. & Quevedo, F. 1988 *Nucl. Phys.* B **301**, 157.
Inoue, K., Kakuto, A., Komatsu, H. & Takeshita, S. 1982 *Prog. theor. Phys.* **68**, 927.
Inoue, K., Kakuto, A., Komatsu, H. & Takeshita, S. 1984 *Prog. theor. Phys.* **71**, 413.
Kawai, H., Lewellen, D. C. & Tye, S. H. H. 1986 *Phys. Rev. Lett.* **47**, 1832.
Kawai, H., Lewellen, D. C. & Tye, S. H. H. 1987 *Nucl. Phys.* B **288**, 1.
Kounnas, C., Nanopoulos, D. V., Srednicki, M. & Quiros, M. 1983 *Phys. Lett.* B **127**, 82.
Lazarides, G., Panagiotakopoulos, C. & Shafi, Q. 1986 *Phys. Rev. Lett.* **56**, 432.
Leontaris, G. 1988 CERN preprint TH.4986/88.
Leontaris, G. & Nanopoulos, D. V. 1988 Madison-CERN preprint MAD/TH/88-16 and CERN-TH.5075.
Lerche, W., Lüst, D. & Schellekens, A. N. 1987 *Nucl. Phys.* B **287**, 477.
Lutken, C. A. & Ross, G. G. 1988 CERN preprint CERN-TH.5132/88.
Lutken, C. A. & Ross, G. G. 1989 (In preparation.)
Mangano, M. 1985 *Z. Phys.* C **28**, 613.
Masiero, A., Nanopoulos, D. V., Tamvakis, K. & Yanagida, T. 1982 *Phys. Lett.* B **115**, 380.
Müller, M. & Witten, E. 1986 *Phys. Lett.* B **182**, 28.
Nandi, S. & Sarkar, U. 1986 *Phys. Rev. Lett.* **56**, 564.

Narain, K. S. 1986 *Phys. Lett.* B **169**, 41.
Narain, K. S., Sarmardi, M. H. & Vafa, C. 1986 Harvard preprint HUTP-86/A089.
Narain, K. S., Sarmardi, M. H. & Witten, E. 1987 *Nucl. Phys.* B **279**, 369.
Ross, G. G. 1988*a* *Phys. Lett.* **200**, 441.
Ross, G. G. 1988*b* CERN preprint, CERN-TH.4935/88.
Sakai, N. & Yanagida, T. 1982 *Nucl. Phys.* B **197**, 533.
Schellekens, A. N. 1987 CERN preprint TH.4807.
Schimmrigk, R. 1987 *Phys. Lett.* B **193**, 175.
Sotkou, G. & Stanishkov, M. 1988 SISSA/ISAS preprint 107EP.
't Hooft, G. (ed.) 1980 *Proc. Advanced Study Inst., Cargese, 1979*. New York: Plenum Press.
Weinberg, S. 1982 *Phys. Rev.* D **26**, 287.
Witten, E. 1986 *Nucl. Phys.* B **268**, 79.
Yamagishi, H. 1983 *Nucl. Phys.* B **216**, 508.
Yau, S. T. 1985 In *Proc. Argonne Symp. anomalies, geometry and Topology* (ed. W. A. Bardeen & R. A. White). Singapore: World Scientific.

Discussion

J. R. Ellis, F.R.S. (*Theory Division, CERN, Switzerland*). It seems to me that supersymmetry breaking is one of the most important open questions in superstring phenomenology. In particular, gaugino condensation alone seems to be inadequate, as it gives a potential which is unstable except when the gauge coupling vanishes. Does Dr Ross have any comment on this?

G. G. Ross. I agree that supersymmetry breaking is of great importance in determining the low-energy structure of the theory. Although gaugino condensation by itself is inadequate the inclusion of (string world-sheet) non-perturbative effects can give a potential with a stable minimum for non-vanishing gauge coupling. A model calculation suggests this has the structure needed to generate a large hierarchy of masses (Ross 1988*a*).

J. R. Ellis, F.R.S. I have predictions for m_b/m and for m_t. Does Dr Ross have any such predictions?

G. G. Ross. The predictions for quark masses may be obtained in the manner detailed in §6.7, but the detailed renormalization group analysis needed to determine these predictions has still to be performed.

Phil. Trans. R. Soc. Lond. A **329**, 395–399 (1989)

Printed in Great Britain

Third quantization

By A. Strominger

Department of Physics, University of California, Santa Barbara, California 93106, *U.S.A.*

The problem of describing a system of interacting four-dimensional universes is addressed. String theory provides a soluble model of interacting two-dimensional universes, and suggests a 'third-quantized' description of the four-dimensional problem. Some fascinating consequences of the existence of other universes are discussed within this framework.

It is logically possible, if not intuitively plausible, that the topology of space-time might fluctuate at very short distances. In quantum mechanics almost everything is uncertain, so why should the topology of space-time be absolutely fixed?

String theory is the only known potentially consistent quantum theory of gravity, so it is natural to try to use it to address this question. Space-time topology is fixed in string perturbation theory, but our understanding of non-perturbative string theory is unfortunately far too rudimentary to address the question of whether or not topology fluctuates. In fact, this is an example of a question to which string theory may not provide an answer, even in principle. String theory may be consistent both with and without topology change. Put another way, there are new coupling constants governing the strength of topological fluctuations, and there is no known argument that string theory fixes them.

String theory has, however, been useful in quite another manner in addressing this question. The many possible types of topological fluctuations include a change in the number of connected components of space. This leads us to consider an interacting many-universe system, and there are severe conceptual and practical difficulties in describing such a system. Viewed as a system of interacting two-dimensional universes, string theory provides us with a soluble model of such a system.

In the past year and a half there has been a lot of progress in understanding the consequences of this type of topology change (if it does occur), in part as a result of acquired wisdom from string theory. (For extensive references, and a recent review, see Strominger (1988).) As we shall see, the consequences are quite remarkable and are even potentially observable.

Our starting point for describing topological fluctuations is the euclidean sum-over-four-geometries weighted by the Einstein action. We will be particularly interested in the process illustrated in figure 1 of the nucleation (or annihilation) of a small 'baby' universe by a large 'parent' universe (Strominger 1984; Hawking 1987, 1988; Lavrelashvili *et al.* 1987, 1988; Giddings & Strominger 1988 *a*). The rate of this process is obtained in a manner (to be described) from the weighted sum-over-four-geometries with appropriately fixed three geometries on the boundaries.

There are a number of obstacles to evaluating this sum. One problem is that Einstein gravity is non-renormalizable. However, the problem of describing topology fluctuations can in some cases be untangled from the short-distance problems of quantum gravity. This is accomplished

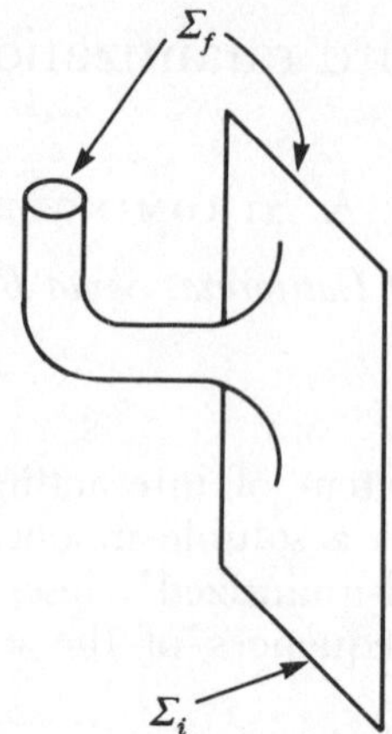

FIGURE 1. A tunnelling process in which the topology of space changes from R^3 on Σ_i to $R^3 \otimes S^3$ on Σ_f.

by regulating the ultraviolet divergences with a short-distance cut-off a, and restricting our attention to cut-off-independent results. This of course will only allow a description of topological fluctuations large relative to the cut-off a.

Another problem is that the euclidean Einstein action is indefinite. This implies that even the regulated functional integral is not convergent. This issue is of great importance, as it has been argued that this divergence explains why the cosmological constant vanishes (Baum 1983; Hawking 1984; Coleman 1988a), but it unfortunately is not well understood at present. It has an analogue in string theory, in that the lorentzian signature of space-time leads to a divergent euclidean world-sheet functional integral. We will assume that here, as in string theory, analytic continuation can be used to define the functional integral.

Even with these assumptions, one cannot of course evaluate the functional integral exactly. What is needed is a small expansion parameter on which to base an approximation scheme. Such a parameter has recently been found (the Peccei–Quinn scale divided by the Planck scale), along with an instanton around which to expand, in a theory of gravity coupled to axions (Giddings & Strominger 1988a). The intractable sum-over-four-geometries is then approximated by a tractable sum-over-instantons. In practice, the expansion around these instantons is our definition of the functional integral describing topology change.

The discovery of these instantons has led to much more concrete discussions of the consequences of baby universes and topology change than was previously possible. One important general feature that has emerged is that baby universes shift space-time coupling constants (Coleman 1988b; Giddings & Strominger 1988b). To see this, consider a small baby universe which contains a positron–electron pair and a photon. The process of it joining a parent universe is illustrated in figure 2a.

It emerges from the instanton calculation that such processes are very unlikely unless the baby universe is very small, of order the Planck length. An observer who cannot make measurements on such small length scales will mistake this process for that of figure 2b. Figure 2b corresponds to a $\bar{\psi}A\psi$ coupling, and a renormalization of the fine structure constant. For example, in a world with no fundamental $\bar{\psi}A\psi$ coupling, one would be induced by the baby universes. Baby universes in general will shift all coupling constants.

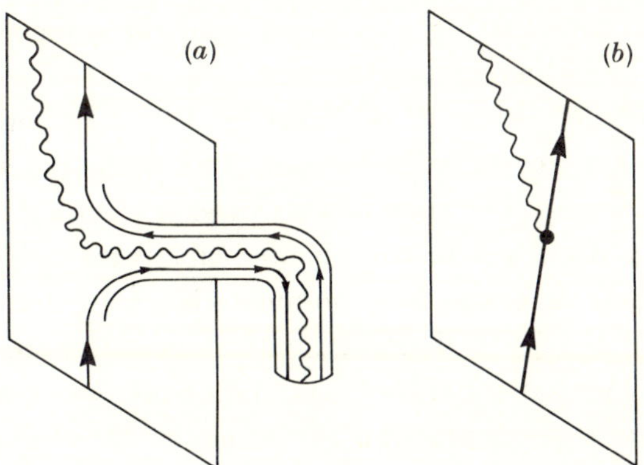

FIGURE 2. (a) Interaction with a baby universe carrying a positron, electron (solid lines with arrows) and a photon (wavy line) is indistinguishable at long distances on the parent universe from (b) a local $\bar{\psi}A\psi$ coupling.

Let us now be a little more precise about this. The hamiltonian for photons and electrons in the parent universe may be written as

$$H = H_0 + eH_1 + \Delta H_1,$$

where H_0 is the free hamiltonian and

$$H_1 = \int \mathrm{d}^3x\, \bar{\psi}A\psi.$$

ΔH_1 represents the effects from the process of figure $2a$. These may be described by the introduction of an operator Φ which adds or substracts a baby universe with an electron, a positron and a photon from the bath of baby universes floating around in the void. Φ is referred to as a 'third-quantized' field operator because it creates or annihilates entire second-quantized single-universe states. One then has

$$\Delta H_1 = \Phi H_1.$$

Now suppose (Coleman 1988 b) that the baby universes are in a coherent state $|\beta\rangle$ such that

$$\Phi|\beta\rangle = \beta|\beta\rangle$$

for some c-number β. The operator Φ in H may then be replaced by its eigenvalue β;

$$H = H_0 + (e+\beta)\,H_1.$$

In this case the entire effect of the baby universes is simply to shift the fine structure constant by an amount which depends on the state of the baby universes. This is an example of the important general result

second-quantized coupling constant = third-quantized field eigenvalue.

This result sounds rather strange in this context, but it is actually quite familiar in string theory. It is just the statement that the coupling constants of the two-dimensional string universe are the space time fields of string field theory

In string theory, if the two-dimensional coupling constants do not take special values, there will be tadpoles and divergences in the scattering amplitudes. The second-quantized string field will condense and shift the two-dimensional coupling constants so that they obey the string equation of motion. This results in an equation of motion, or dynamical constraint on coupling constants in the two-dimensional string universe. Could the same be true for baby universe induced couplings in our four-dimensional universe?

Before this question can be answered, we must give a more complete description of the dynamics of an interacting many-universe system. The basic process of figure 1 obviously cannot be considered in isolation. Iteration of that process, and accounting for the possibility of baby universes growing up, leads to complicated diagrams as depicted in figure 3. The proper interpretation of these diagrams is by no means obvious. Because each universe has its own time, we certainly can not define dynamics with respect to time on a single universe. We instead proceed by analogy with string theory, and define the system as a quantum field theory on superspace, the space of three geometries (Giddings & Strominger 1988c; Banks 1988). (Alternate interpretations are discussed by Hawking & Hartle (1983) and Coleman (1988a).) The third-quantized field is a function on superspace, and acts on the 'void' to create entire second-quantized states of a single universe.

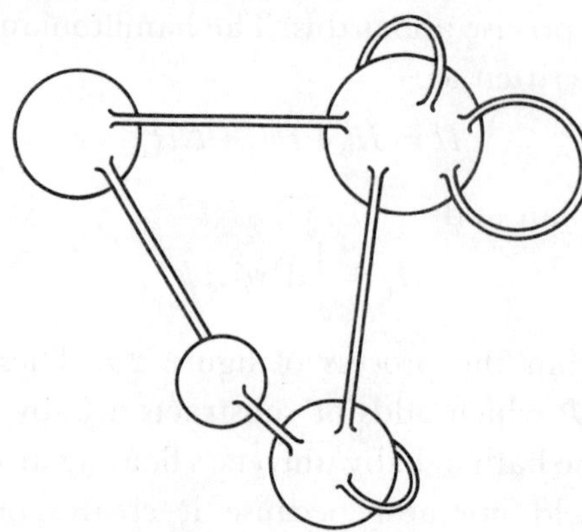

FIGURE 3. In general there can be many parent universes (large spheres) connected in many ways by baby universes (thin tubes).

The action of this third-quantized field theory – a functional of functions on superspace – is defined as the action whose Feynman diagrams reproduce the second-quantized sum-over-four-geometries. This is exactly the route by which correlation functions on the string world sheet are reinterpreted as space-time scattering amplitudes. The analogy is displayed in table 1.

In practice, the third-quantized action has not been constructed. However, in the instanton approximation, the action need only reproduce the sum-over-four-instantons. Rather than quantum field theory on superspace, one then has only quantum field theory on the moduli space of the instanton! In the model of Giddings & Strominger (1988c) this space is the real line. The resultant quantum mechanics model can then be constructed quite explicitly.

TABLE 1

string field theory	third-quantized field theory
string	universe
string field	third-quantized field
loop space	superspace of three geometries
$L_0 + \tilde{L}_0 - 2$	Wheeler–De Witt operator
first-quantized BRST charge	second-quantized BRST charge
vacuum	void

In this framework, the third-quantized field, or equivalently the second-quantized space-time couplings, are subject to the third-quantized equation of motion. This constrains the values of the space-time couplings in a manner which would appear quite mysterious to the innocent four-dimensional observer. In the model of Giddings & Strominger (1988 c), one of these constraints is the vanishing of the cosmological constant (Coleman 1988 a). Thus the existence of other universes is not only fascinating conceptually, it even has a chance of explaining facets of the universe we inhabit.

This work was supported in part by DOE Outstanding Junior Investigator Grant DE-AT03-76ER70023 and an A. P. Sloan Foundation Fellowship, BR-2623T.

REFERENCES

Banks, T. 1988 Prolegomena to a theory of bifurcating universes: a nonlocal solution to the cosmological constant problem, or little lambda goes back to the future. Santa Cruz preprint SCIPP 88/09.

Baum, E. 1983 *Phys. Lett.* B **133**, 185.

Coleman, S. 1988 a Why there is nothing rather than something: a theory of the cosmological constant. Harvard preprint HUTP-88/A022.

Coleman, S. 1988 b Black holes as red herrings: topological fluctuations and the loss of quantum coherence. *Nucl. Phys.* B **307**, 864.

Giddings, S. B. & Strominger, A. 1988 a Axion-induced topology change in quantum gravity and string theory. *Nucl. Phys.* B **306**, 890.

Giddings, S. B. & Strominger, A. 1988 b Loss of incoherence and determination of coupling constants in quantum gravity. *Nucl. Phys.* B **307**, 854.

Giddings, S. B. & Strominger, A. 1988 c Baby universes, third quantization and the cosmological constant. Harvard preprint, HU TP-88/A036.

Hawking, S. W. 1984 The cosmological constant is probably zero. *Phys. Lett.* B **134**, 403.

Hawking, S. W. 1987 Coherence down the wormhole. *Phys. Lett.* B **195**, 337.

Hawking, S. W. 1988 Wormholes in spacetime. *Phys. Rev.* D **37**, 904.

Hawking, S. W. & Hartle, J. B. 1983 *Phys. Rev.* D **28**, 2960.

Lavrelashvili, G. V., Rubakov, V. A. & Tinyakov, P. G. 1988 *JETP Lett.* **46**, 167.

Strominger, A. 1984 Vacuum topology and incoherence in quantum gravity. *Phys. Rev. Lett.* **52**, 1733.

Strominger, A. 1988 Baby universes, UCSB preprint; to appear proceedings 1988 TASI summer school.

Discussion

J. R. ELLIS, F.R.S. (*Theory Division, CERN, Geneva, Switzerland*). Professor Strominger mentioned recent ideas that may lead to a solution of the cosmological constant problem. However, the same ideas seem to lead to problems elsewhere, e.g. the masses of spin zero particles, such as the pion, should vanish. Is it possible to avoid throwing out the baby with the bathwater?

A. STROMINGER. It is true that it has been argued by some that the logic that leads to the vanishing of the cosmological constant also leads to the vanishing of the pion mass (while others argue it does not). More generally, there is a very strong singularity that forces the cosmological constant to zero, and there is a danger that all the constants of nature will be dragged to zero (or infinity) along with it. However, what has been done so far should really just be regarded as a first stab at calculating the constants of nature. There are many suspicious assumptions in the calculations of the cosmological constant, and many more in calculations of the pion mass. The physics community is now in the process of scrutinizing these assumptions. We hope to be able to answer this question soon, one way or the other.

Phil. Trans.¨R. Soc. Lond. A **329**, 401–413 (1989)

Printed in Great Britain

Strings at superplanckian energies: in search of the string symmetry

By D. J. Gross

Joseph Henry Laboratories, Princeton University, Princeton, New Jersey 08544, U.S.A.

The characteristic energy scale of superstring theory, which attempts to unify all the interactions of matter with gravity, is the Planck energy of 10^{28} eV. Although this energy is 16 orders of magnitude higher than currently accessible energies, it is important to consider the nature of string physics in this region since it could shed light on the non-perturbative physics at the Planck scale, which determines the structure of the vacuum.

In this paper I review some recent attempts to explore this domain. In particular, I discuss string scattering at very high energies, the indications of the existence of a large symmetry that is restored at short distances and the possible breakdown of our concepts of space-time at these energies.

1. Introduction

We have heard much in this meeting about the technology of two-dimensional conformal field theory, which is useful for the construction of classical vacua for string theory. However, the real problem in string theory is not to find more and more classical solutions, but to find the dynamics that picks the unique vacuum, and thereby to make contact with the real world.

It is not unusual to have more than one classical vacuum in a quantum mechanical system, especially in supersymmetric theories where there are often many 'flat directions' in which one can vary the expectation value of some field without changing the energy of the vacuum. In such a situation one can develop a perturbation theory about each vacuum separately. However, unless perturbative instabilities arise, the correct vacuum involves non-perturbative physics, e.g. tunnelling. The trouble is that in string theory we have a similar situation, but we do not know the lagrangian, we do not even know what the coordinates are. All we have are the classical solutions and perturbation theory about them. All of our understanding of string theory is based on the following two ingredients: a rule for finding the classical vacua (equivalent to classical solutions) of some theory whose lagrangian we do not know, and rules for calculating S-matrix elements in perturbation theory.

This is not good enough. One of the reasons it is not good enough is that perturbation theory diverges very badly. This is a recent result of Gross & Periwal (1988) (our proof is for the bosonic string theory, mainly for technical reasons, but I believe it is true for all string theories). That is, if you construct string scattering amplitudes perturbatively in powers of the coupling (which in principle is determined by the dynamics but in practice, namely in perturbation theory, is a free parameter), then you discover that this perturbation expansion has zero radius of convergence. The coefficients grow faster than $n!$, so not only is it divergent it is also not Borel summable. This is bad for perturbation theory but very good for physics. If it were not the case that perturbation theory diverged then you could sum it, ending up with billions and billions of classical vacua, and the full quantum mechanical perturbations about them. Because these are all consistent theories, order by order in the coupling, you would end up with billions of

[83]

consistent theories. All of them would contradict observation as all of them have unbroken supersymmetry and all of them have a massless dilaton. So it is good that we are not allowed to sum the series.

The meaning of a perturbation theory that diverges in this way is that the vacuum is unstable. One indication of this is that if you try to sum such a pertubative expansion (say by the Borel technique), you would find that the vacuum has complex energy, which indicates vacuum instability. Such series occur often in quantum mechanics, for example in the case of the double well potential. They also occur in totally consistent physical field theories, such as quantum chromodynamics (QCD). In both of these cases the non-Borel summable divergences are related to the existence of instantons, which are signals of quantum tunnelling that mixes different vacua. Tunnelling effects behave as $\exp[-1/g^2]$, which cannot be reproduced by a perturbative expansion. We regard this divergence as an indication that all of the string solutions that people have found are unstable at the quantum level as a result of non-perturbative effects. We must develop a formalism that allows us to go beyond perturbation theory, find the non-perturbative effects that will pick out a unique vacuum, break supersymmetry, generate a mass for the dilaton and all other good things.

The main trouble in going beyond perturbation theory is that we do not have the analogue of the lagrangian, or a second-quantized hamiltonian. We do not even know what the natural variables of string theory are, nor what the natural description of configuration space is. All we have are these rules for constructing S-matrix amplitudes in perturbation theory. There have been many attempts to go beyond perturbation theory to find the correct basic formulation of the theory. I think it honest to say that none of these suggestions has got very far. A sign of how bad things are is the difficulty in finding non-stationary classical solutions of string theory. An appropriate *euclidean* solution of this variety could be used to explore non-perturbative phenomena semi-classically. Although we have developed much understanding of stationary classical solutions, which are given by conformal field theories, not one non-stationary solution has been constructed!

I think it is very important to keep an open mind about these issues. The most obvious approach such as the construction of *string field theory* might not be correct. It is very instructive to recall the other instance in which strings appear in physics, namely as a description of the large-scale behaviour of QCD, where they describe the approximate structure of hadronic electric flux tubes. The large-distance behaviour of SU_N gauge theory, in the limit of large N, might very well be described by an effective string theory. To leading order in $1/N$ we would have a non-interacting system of strings, whose spectrum would consist of stable hadrons. The coupling of the hadrons would be proportional to $1/\sqrt{N}$, and one could develop a perturbative expansion of hadronic scattering amplitudes, much as one does in string theory.

If one had such a formulation of QCD it would be very difficult to extend it to a non-perturbative framework. Much of the underlying simplicity, especially the conceptual simplicity, of the theory would not be evident. An analogue of the problems we now face would be the attempt to discover quarks, gluons and non-abelian gauge theory starting from an effective string theory of QCD. The reason is that the formulation of the theory appropriate for large-scale physics does not reveal the essence of QCD. In QCD, as in most theories, the ultimate simplicity lies at short distances. In this régime the variables in terms of which one describes the system are fewer and simpler and the symmetries of the theory are manifest and evident.

If this is true for strings as well, then to make progress we must try to understand the high-energy behaviour, or the high-energy phase, of string theory. We must also be ready for the possibility that a totally different kind of structure will be appropriate for the description of this domain; but we may hope that it will prove to be conceptually simpler and more elegant.

Thus we must explore the physics of strings at energies above their characteristic energy scale. In the case of superstring theory, which attempts to unify all the interactions of matter with gravity, this is the Planck energy of 10^{28} eV. Although this energy is 16 orders of magnitude higher than currently accessible energies, it is important to study this domain. The idea is to do what an experimental high-energy physicist would if she had unlimited funds available, namely build an accelerator with the highest possible energy and scatter protons (or gravitons) and look at what comes out. We don't have unlimited funds but we have a theory, or at least a small handle on the theory. So we can do thought (gedanken) experiments, pushing the theory to its limits where it might reveal new structures, or at least new problems. In particular, if the theory possesses a much larger symmetry than the symmetry that is evident in low-energy scattering amplitudes (i.e. supergauge + general coordinate invariance) than it might become evident in the structure of high-energy scattering amplitudes. In this paper I review some recent attempts to explore this domain. In particular I discuss string scattering at very high energies, the indications of the existence of a large symmetry that is restored at short distances and the possible breakdown of our concepts of space-time at these energies.

2. The high-energy behaviour of string scattering amplitudes

I shall describe, briefly, the study that I carried out, with Paul Mende (Gross & Mende 1987, 1988) of the high-energy behaviour of string scattering. From these calculations a few very interesting features emerge. First, the limit of very high energies or, because the only mass scale in the theory is the Planck mass, the zero Planck mass limit, is the *semi-classical* limit of first-quantized string theory. Namely, order by order in the perturbative expansion of string scattering amplitudes, the functional integrals over the space-time surfaces mapped out by the propagating strings are dominated by a saddle point, which corresponds to a classical solution of the string equations of motion on a particular Riemann surface.

The fact that the zero Planck mass limit is a semi-classical limit has interesting implications that I shall discuss. One of them is that it is easy to calculate the high-energy behaviour order by order in perturbation theory. Also, the result has a large amount of universality and is very stringy and unusual. Finally, there appears in the structure of the scattering amplitudes hints of a large and totally mysterious symmetry of string theory that is restored as we take the Planck mass to zero (Gross 1988).

Let us first examine the behaviour of string scattering at large energy in the Born approximation. This is straightforward because the amplitude is given as a ratio of gamma functions and the limit is easily deduced with the aid of Stirling's formula. Consider the elastic scattering of four tachyons, the ground state of the bosonic string. (For simplicity I shall discuss mostly the bosonic string, although most of my remarks hold for the super and heterotic strings as well.) The Virasoro–Shapiro amplitude is given by

$$A_{\text{tree}} = g^2 \frac{\Gamma(-1-\tfrac{1}{8}s)\,\Gamma(-1-\tfrac{1}{8}t)\,\Gamma(-1-\tfrac{1}{8}u)}{\Gamma(2+\tfrac{1}{8}s)\,\Gamma(2+\tfrac{1}{8}t)\,\Gamma(2+\tfrac{1}{8}u)}, \tag{2.1}$$

28-2

where s, t, u are the Mandelstam variables $s = -(p_1+p_2)^2$, $t = -(p_1+p_3)^2$, $u = -(p_1+p_4)^2$; $p_i^2 = 8$ and $s+t+u = -32$. For large energy, $s \to \infty$, and fixed s-channel scattering angle ϕ, $\sin^2 \frac{1}{2}\phi = -t/(s+32) \approx -t/s$, $\cos^2 \frac{1}{2}\phi = -u/(s+32) \approx -u/s$, we have,

$$A_{\text{tree}} = 8ig^2 \, e^{-8}(stu)^{-3} \exp\left[-\tfrac{1}{4}(s \ln s + t \ln t + u \ln u)\right]$$

$$= ig^2 2^9 s^{-1}(\sin \phi)^{-6} \exp\{\tfrac{1}{4}(s+32)\,[\sin^2 \tfrac{1}{2}\phi \ln \sin^2 \tfrac{1}{2}\phi + \cos^2 \tfrac{1}{2}\phi \ln \cos^2 \tfrac{1}{2}\phi]\}. \tag{2.2}$$

This exponential fall-off is quite different from that of the power behaviour characteristic of field theory. It is clearly an essential feature of string theory, surely related to the finiteness of this theory, yet its deeper meaning is still unclear. It contradicts the rigorous lower bound of quantum field theory, which states that $|A(s, \cos \phi)| \geqslant \exp[-\sqrt{s} \ln s f(\phi)]$.

The proof of this bound uses unitarity, the existence of a finite mass gap and, most importantly, the assumption of polynomial boundedness in the energy, for fixed t, of the scattering amplitudes. Of course, perturbative string theory amplitudes violate all of these assumptions so there is no contradiction. In quantum field theory polynomial boundedness is a consequence of locality, and the power fall-off at high energies is a consequence of power singularities in the operator product of local fields at light-like separations. In the days of axiomatic field theory people asked themselves what is the worst kind of singularity that one could tolerate in the product of field operators, $\phi(x)\phi(y)$, and still have a local theory. For example, if you have any finite sum of delta functions, $(\partial_x^n \delta(x-y))$, than you get a local distribution, which vanishes when $x \neq y$. But you can even have an infinite sum as long as the coefficients do not grow too rapidly. When you work out what that means in momentum space it means that you can't have more raid fall-off than $\exp(-\sqrt{s})$. If you violate the lower bound you violate locality.

In string theory we find this crazy exponential behaviour, crazy from the point of view of field theory, and therefore we should be interested because it is a sign that something new is happening at short distances in string theory. There certainly is no way that one could represent this physics by an effective field theory that is truly local. This is one of the most interesting lessons that we learn from the calculation of high-energy string scattering.

Let us now consider the general G-loop amplitude, $A_G(P_i)$, given by the path integral

$$A_G(P_i) = \int \frac{\mathscr{D} g_{\alpha\beta}}{\mathscr{N}} \mathscr{D} X^\mu \exp\left\{-\frac{1}{2\alpha'} \int \mathrm{d}^2\xi \sqrt{g}\, g^{\alpha\beta} \partial_\alpha X^\mu \partial_\beta X_\mu\right\} \Pi\, V_i(P_i), \tag{2.3}$$

where the string tension α' is proportional to M_{Planck}^{-2}, V_i is the vertex operator for the emission of the ith particle and the integral is over all compact surfaces of genus G. In this formula it looks like α' plays the role of Planck's constant and $\alpha' \to 0$ is the classical limit. However, we must remember that to discuss the classical limit of a theory one must specify the observables. In our case the observables are the vertex operators of physical states, all of which depend on the external momenta as

$$V(P_i) = \int \mathrm{d}^2\xi_i \sqrt{g}\, e^{iP_i X(\xi_i)} \mathscr{V}(P_i, X(\xi_i)),$$

where $\mathscr{V}(P_i, X(\xi_i))$ depends on the quantum numbers of the particle. In the limit of high energy, or $\alpha' \to \infty$, or $M_{\text{Planck}} \to 0$, it is then clear that $X^\mu \approx \alpha'(1/\partial^2)\,P^\mu$. In other words X^μ should be scaled by α', in which case the exponent appears with α' upstairs. Thus $\alpha' \to \infty$, or $M_{\text{Planck}} \to 0$, is the semi-classical limit and in this limit the integral will be dominated by a particular two-dimensional surface, X^μ and $g_{\alpha\beta}$. In the critical dimension the integral is invariant under the group of diffeomorphisms and Weyl rescalings of the metric, which is why we have divided by $\mathcal{N}$, the volume of this group. It can then be reduced to an integral over a slice of the finite-dimensional moduli space $\mathcal{M}_G$ of Riemann surface of genus G (D'Hoker & Phong 1989). Consider, for simplicity, external tachyons whose vertex operators are $\mathcal{V}(P_i) = 1$. The X^μ integral is easily done yielding

$$A_G = g^{2G+2} \int_{\mathcal{M}_G} [\mathrm{d}\boldsymbol{m}] \prod_i \mathrm{d}^2\xi_i \sqrt{g(\xi_i)}\, \Omega(\boldsymbol{m}, \xi_i) \exp\left[-\tfrac{1}{2}\sum P_i P_j G_m(\xi_i, \xi_j)\right], \qquad (2.4)$$

where $\boldsymbol{m}$ are a set of coordinates for moduli space, $\Omega(\hat{\boldsymbol{m}}, \hat{\xi}_i)$ is the measure on the moduli space and $G_m(\xi_i, \xi_j)$ is the scalar Green function, $\Delta_0 G_m(\xi_i, \xi_j) = -2\pi\delta^2(\xi_i, \xi_j)$, on the surface.

The only place that the momenta enter into this expression is in the exponential. Therefore, in the limit that *all* $P_i P_j$ become large, we can perform the integral by saddle-point techniques. The exponent,

$$\mathcal{E}(P_i, \xi_i, \boldsymbol{m}) \equiv \tfrac{1}{2}\sum_{i<j} P_i P_j G_m(\xi_i, \xi_j), \qquad (2.5)$$

can be regarded as the electrostatic energy of two-dimensional Minkowski charges P_i, placed at positions ξ_i on a Riemann surface of genus G with moduli $\boldsymbol{m}$. Thus to determine the amplitude, we must find the values of ξ_i and $\boldsymbol{m}$ for which the action $\mathcal{E}(P_i, \xi_i, \boldsymbol{m})$ is extremal. This is equivalent to the problem of finding the points of electrostatic equilibrium (in general unstable) on an arbitrary two-dimensional surface which is allowed to change its shape at no energy cost! This is a beautiful variational problem. It does not appear to have been investigated, perhaps because it is only for Minkowski charges (where we can set $P_i^2 \approx 0$ compared to $P_i P_j$) that the problem is strictly conformally invariant.

Because this exponential term is common to all string theories, as well as to amplitudes involving other external states, *the dominant large-energy behaviour of string amplitudes will be independent both of the theory and the particular quantum numbers of the scattered particles*. It will also be independent of the nature of the compactification of some of the spatial dimensions, as long as we consider scattering particles that are not wrapped many times around some internal dimension or whose internal momenta are not large. Of course the prefactor will depend to some degree on the theory and on the external particles. This is an important result. It is an indication of the universality of high-energy behaviour, analogous to the universality of the singularities of the operator product expansion in field theory. It opens the way for the systematic study of the high-energy behaviour of amplitudes with any number of external particles of any type. Such an investigation could throw light on what replaces the operator product expansion in string theory.

This behaviour also means that the determination of the asymptotic behaviour of fixed angle, high-energy scattering in string theory is a much simpler problem than in field theory. If we were to try to evaluate the behaviour of field theoretic amplitudes by similar methods (i.e. introduce Schwinger or Feynman parameters, perform the integrals over the momenta and search for a saddle point in the remaining integrals), we would fail. The integrals over the Feynman or Schwinger parameters, the analogue of the moduli for field theory, are dominated by end-point contributions. One can use power counting to estimate the fall-off of field theory graphs in perturbation theory, but the precise coefficients are not easily obtained. In string theory it turns out, as we shall see below, that the saddle points lie in the middle of moduli space, far from the boundaries. For this reason we can ignore the infrared divergences of the bosonic string. They arise from points on the boundary of moduli space which do not dominate the high-energy behaviour. In other words, if we were to cut off the infrared instabilities of the bosonic theory by means of a cut-off, the high-energy behaviour would be independent of the cut-off.

In string theory, we will be able to derive a precise form for the asymptotic behaviour of the G-loop amplitude. Not only can we easily find the dominant saddle point, $(\hat{m}, \hat{\xi}_i)$, but we will be able to calculate the measure on moduli space at these points, $\Omega(\hat{m}, \hat{\xi}_i)$, as well as calculate the gaussian fluctuations about the saddle!

Let us redo the tree approximation by considering electrostatics on a sphere. Take a sphere, put four Minkowski charges (whose sum is zero) on the sphere and find the positions of equilibrium. The sphere is conformally the same as the complex plane (including the point at infinity), as can be seen by a stereographic projection that is a conformal transformation. The Green's function on the plane is simply $G(z, z') = \ln|z - z'|$. So the electrostatic energy is very simple. It is just

$$\mathscr{E} = s[\ln|z_1 - z_2| + \ln|z_3 - z_4|] + t[\ln|z_1 - z_3| + \ln|z_2 - z_4|] + u[\ln|z_1 - z_4| + \ln|z_2 - z_3|].$$

Because all spheres are equivalent to the complex plane there are no moduli. So the only thing we have to vary are the positions of the charges. Because of the conformal invariance this problem has an $\mathrm{SL}(2, C)$ (Moibus) invariance. The action will be invariant under a transformation $z \to (az + b)/(cz + d)$, that takes the plane into the plane. The only $\mathrm{SL}(2, C)$ invariant combination of four zs is their cross ratio,

$$\lambda = (z_1 - z_3)(z_2 - z_4)/(z_1 - z_4)(z_2 - z_3).$$

By using momentum conservation it is trivial to see that, because $s + t + u = 0$,

$$\mathscr{E} = t \ln|\lambda| + u \ln|1 - \lambda|,$$

which is easy to extremize. Differentiating with respect to λ we find $t/\lambda = u/(1 - \lambda)$. Thus $\lambda = -t/s = \sin^2 \frac{1}{2}\phi$. Plugging this back into the action we find that the minimal action is just the expression we derived before by using Stirling's formula. So we reproduce the result for the sphere and we have understood that this expression comes from a saddle point. Similar methods can then be used to discuss an arbitrary Riemann surface with any number of handles.

It turns out that to study the problem of a Riemann surface of large genus it is useful to think of a Riemann surface as a branched cover of the complex plane, or as an *algebraic curve*, $y = f(z)$, where y and z are complex variables. This describes a curve in CP^2, or equivalently y is a meromorphic function of z and describes a branched cover of the complex plane, a Riemann surface.

For example the torus is a two-sheeted cover of CP^1, the complex plane including the point at infinity, branched over four points; i.e. the algebraic curve

$$y^2 = (z - a_1)(z - a_2)/(z - a_3)(z - a_4). \tag{2.6}$$

To construct the torus in this way, take two copies of the complex plane, one over the other. Next, make square root branch cuts on each sheet between a_1 and a_3, and between a_2 and a_4. Then construct 'crossed bridges', identifying the left edge of a cut on one sheet with the right edge of the corresponding cut on the other, and vice versa. The surface is clearly closed and connected, and a simple triangulation shows it has genus 1. What we have done is connect two spheres by bridges to obtain a torus.

We are interested in an algebraic curve of the same type, an N sheeted cover of the Riemann surface. This is a curve with branch points at a_i, each one of which has N sheets, so that when you go around each of these branch points N times you come back to the original sheet. The most general curve of this form is

$$y^N = \prod_{i=1}^{L} (z - a_i)^{L_i}. \tag{2.7}$$

As long as the L_is are relatively prime to N then each branch point a_i is of order $N-1$ and is a point common to all N sheets. (Note that a square root branch point is of order one!) The number of branch points is precisely L if we impose that $\sum L_i = 0 \pmod{N}$, so that the point at infinity is regular. The curve has a Z_N automorphism, the largest generic automorphism of a Riemann surface. It is not difficult to show that the genus of such a curve is given by

$$G = \tfrac{1}{2}(N-1)(L-2). \tag{2.8}$$

I believe that the dominant saddle points, the elastic scattering at high energy, are just such Z_N curves, with four branch points at which we put the four external charges (momenta). A remarkable feature of these curves is that we can easily do electrostatics on them. On a general Riemann surface electrostatics, i.e. calculating the inverse of the laplacian, is a very complicated business. But *electrostatics is trivial on a branched cover of the sphere if the charges are all placed at the branch points*! This is because each sheet is simply a copy of the complex plane, and each sheet sees the same sources because the points where the charges are located are common to all of the sheets. It is easy to see that in this circumstance the electrostatic field will be proportional to the electrostatic field of the four charges on the complex plane and can immediately be written down, no matter how many sheets we have. Let us consider the most general situation of an N-sheeted Riemann surface, where we place charges P_i at the branch points a_i. (There may be more than four branch points and some of the P_i might vanish. We suppress the Minkowski indices.) We claim that the electric field produced by these charges, $E(z) = E_x(z) + iE_y(z)$, take the same value on all N sheets, and is given by

$$E_N(z) = \frac{1}{N} E_1(z) = \frac{1}{N} \sum_i \frac{P_i}{z - a_i}, \tag{2.9}$$

where $E_1(z)$ is the electric field of the trivial genus 0 case, the sphere.

In a similar fashion we can construct the saddle point expression for the space-time coordinate X^μ, which is equal in our analogue electrostatic problem to the electrostatic potential. It is therefore given by

$$X_{cl}^\mu(z) = \frac{i}{N} \sum_{i=1}^{4} P_i^\mu \ln|z - a_i|. \tag{2.10}$$

It is now trivial to calculate the action for the N-sheeted curve considered above. Because

$$\mathscr{E} \propto \int |E_N|^2 = N \int_{\text{sphere}} \left| \frac{1}{N} E_1 \right|^2,$$

it follows that

$$\mathscr{E}_N = \mathscr{E}_1/N = -\frac{1}{2N} \sum_{i<j} P_i P_j \ln|a_i - a_j|. \tag{2.11}$$

In the case of four branch points, and four charges located at these branch points, $\mathscr{E}_N$ is proportional to the energy for the sphere. Extremizing with respect to the modulus $\lambda = (a_1 - a_3)(a_2 - a_4)/(a_1 - a_2)(a_3 - a_4)$, we find the same result as before, $\lambda = -t/s$, so

$$\mathscr{E}_N = (1/4N)(s \ln s + t \ln t + u \ln y). \tag{2.12}$$

One problem that remains open is the extension of these results to the many-particle case. I do not even know how to solve the problem of multiparticle high-energy scattering in the Born approximation. The corresponding *electrostatic problem* is to find the equilibrium positions of M Minkowski null charges, P_i, which sit on the sphere. This problem is very straightforward to formulate, is a very pretty problem, but the general solution is unknown. I present it here as a challenge.

The equations one must solve are the following:

$$\sum_{j \neq i} \frac{P_i P_j}{z_j - z_i} = 0, \quad i = 1, \ldots, M, \tag{2.13}$$

which express the vanishing of the force on the ith particle, located at z_i. This equation can be written in many different ways. For example an equivalent condition is that the electric field $E^\mu(z) = \sum_i P_i^\mu/(z - z_i)$ satisfy

$$E^\mu E_\mu = 0. \tag{2.14}$$

This expresses the fact that the induced metric on the string world-sheet is conformal

$$g_{zz} = \partial_z X_\mu \partial_z X^\mu = 0, \quad g_{\bar{z}\bar{z}} = \partial_{\bar{z}} X_\mu \partial_{\bar{z}} X^\mu = 0. \tag{2.15}$$

I know of one solution to the multicharge problem, which exists whenever all M momenta lie within a four-dimensional subspace. In this case the position of the ith charge is independent of the value of all the other charges and is given by

$$z_i = \frac{P_i^1 + \mathrm{i} P_i^2}{P_i^0 + P_i^3} = \frac{P_i^0 - P_i^3}{P_i^1 - \mathrm{i} P_i^2}. \tag{2.16}$$

This solution has a very nice geometric interpretation, which I will discuss elsewhere (Gross 1989)†. The action for this solution is proportional to $\sum_{i \neq j} P_i P_j \ln(P_i P_j)$. The general solution for M particles, for $M > 5$ when generically the particles lie outside of four dimensions, is unknown.

Calculating the classical action is only the beginning of a semi-classical calculation. One must also calculate the measure on moduli space at these points, $\Omega(\hat{m}, \hat{\xi}_i)$, as well as calculate the gaussian fluctuations about the saddle. Remarkably, using many techniques of conformal field theory, it turns out to be possible to explicitly do these calculations (Gross & Mende 1987). The result is

$$A_G = \mathrm{i}^G 2^{23G+8} \pi^{9G+3} (G+1)^{9G+3} \times g^{2G+2} \, \mathrm{e}^{-(s \ln s + t \ln t + u \ln u)/4(G+1)}$$

$$\times (stu)^{-3G^2+4G+9/3G+3} |\lambda(1-\lambda)|^{-6G^2+14/3G+3}$$

$$\times \prod_{k=1}^{G} [J_k(\lambda) J_k(1-\lambda)]^{-\frac{1}{2}} \left[\left(\sin \frac{\pi k}{N} \right)^{-1} F_k(\lambda) F_k(1-\lambda) \right]^{-13}. \tag{2.17}$$

† I have been informed by David Fairlie that this solution was known to him since 1972.

Here, $G = N-1$, $\lambda = -s/t$, $F_k(\lambda) = F(k/N, 1-k/N; 1; \lambda)$,

$$J_k(\lambda) = C_k[(C_k - \tfrac{1}{2}\lambda)(C_k - \tfrac{1}{2}(1-\lambda) - (\tfrac{1}{2} - k/N)^2 \lambda(1-\lambda)], \text{ and } C_k = \lambda(1-\lambda)(\mathrm{d}/\mathrm{d}\lambda) F_k(\lambda).$$

Note that this series is very divergent. This makes it unreasonable to try to sum it. Instead we should try, as I shall below, to deduce properties of the series that are true to all orders.

3. High-energy symmetries

It is often the case that spontaneously broken symmetries of a physical theory are hard to recognize at low energy, but become evident in the high-energy behaviour of the theory. Thus the broken $SU_2 \times U_1$ symmetry of the electroweak interactions can be seen by examining weak scattering amplitudes at energies high enough that the W and Z_0 masses can be neglected. String theory surely possesses a very rich symmetry, as suggested by its incredible degree of uniqueness; however, this symmetry is little understood. Presumably this is because most of the string symmetry is spontaneously broken in the known ground states, leaving only the familiar gauge symmetries unbroken and manifest. Perhaps all the string states are gauge particles, but most are massive because of spontaneous symmetry breaking. Perhaps at very high energies, so high that the Planck mass (which in string theory is proportional to the string tension $T = 1/\pi\alpha'$) can be neglected, the full symmetry of string theory is restored. Perhaps we can discover this symmetry (if it exists) by studying string theory in the high energy (or $\alpha' \to \infty$) limit. In particular, if at high energies a larger symmetry is restored there should then exist linear relations between the scattering amplitudes that should be valid order by order in perturbation theory. If so these might be discoverable by analysing the high-energy behaviour of the theory perturbatively.

We can use the results described above to search, and to find, such relations. We are now interested in the dependence of the high-energy scattering amplitudes on the quantum numbers of the external particles. Recall the expression, (2.3), for the amplitude where *the only factor that depends on the nature of the particles being scattered is $\prod_i \mathcal{V}_i$*! Therefore, if we can determine the saddle-point surface, in other words the $\hat{m}_i$, ξ_i and X^μ_{cl}, which only depend on the number of particles participating in the scattering but not on their quantum numbers, then we can immediately deduce a linear relation between any two scattering amplitudes involving the same number of particles with the same momenta. For example, in the case of the scattering of particles a_i and particles b_i

$$A_G^{a_i}(P_i) = \prod_i \mathcal{V}_{a_i}(X^\mu_{cl}, P_i) / \prod_i \mathcal{V}_{b_i}(X^\mu_{cl}, P_i) \, A_G^{b_i}(P_i) \, (1 + O(\alpha')). \tag{3.1}$$

This is an amusing relation, but one of little use by itself as it only relates amplitudes at a given order of perturbation theory. Fortunately, for the dominant saddle points discussed above the dependence of X^μ_{cl} on the order of perturbation theory, G, is trivial. All of these surfaces (including those with different L_i) give rise to the same exponential factor in (2.3), all of them have the same Green's function and for all of them X^μ is given by (2.10).

It is a remarkable fact that the saddle-point surfaces are all identical, except for the scale (which goes as $1/N$), for all the saddle points in each order of perturbation theory. Consequently, the only genus dependence present in equation (3.1) comes from the $(1/N)$s in the X^μ_{cl}s present in the $\mathcal{V}_i$s, namely $\mathcal{V}(X^{\mu N}_{cl}, P) = \mathcal{V}(X^{\mu 1}_{cl}/N, P)$. If all the $\mathcal{V}_{a_i}$s were homogeneous functions of the X^{μ}s of the same degree, then (3.1) would be independent of G. Unfortunately, this is not the case, except for the vertex operators of the simplest string states. However, the factors of $1/N$ can be replaced by derivatives with respect to the momenta. This is because the operator $\mathscr{D} \equiv 2/\alpha' S \sum_i P_i \cdot \partial/\partial P_i$ brings down a factor of N, when acting on $\mathrm{e}^{-\frac{1}{4}\alpha' S}$, where $S = s \ln s + t \ln t + u \ln u$. When acting on the other polynomial terms in the momenta $\mathscr{D}$ is of order $1/\alpha'$. Therefore we can replace the $1/N$ by $\mathscr{D}$. Thus $\mathscr{V}(X^{\mu N}_{cl}, P) = \mathscr{V}(X^{\mu 1}_{cl}\mathscr{D}, P)$, when acting on $A_{N-1}(P_i)$, to leading order in $1/\alpha'$.

Using this trick we can write down linear relations between any two four-particle scattering amplitudes;

$$\prod_i \mathscr{V}_{b_i}(X^{\mu}_{cl}\mathscr{D}, P_i) A^{a_i}_G(P_i) = \prod_i \mathscr{V}_{a_i}(X^{\mu}_{cl}\mathscr{D}, P_i) A^{b_i}_G(P_i)(1 + O(\alpha')), \tag{3.2}$$

which are independent of G! We can now argue that this relation should hold for the complete amplitude, in the $\alpha' \to \infty$ limit, because it holds order by order in perturbation theory. This is a traditional method of proving properties of quantum field theory (such as the operator product expansion, symmetries, the renormalization group, etc.) by establishing their validity order by order in perturbation theory. It is not, however, without dangers, especially in proving asymptotic theorems. None the less we shall make this assumption. Then, because as $\alpha' \to \infty$ all the particles become massless, we can relate the scattering amplitudes for any set of particles. Indeed all the four-particle scattering amplitudes can be expressed in terms of, say, the four-tachyon amplitude $A_{\text{tachyon}}(P_i)$:

$$A_{a_i}(P_i) = \prod_i \mathscr{V}_{a_i}(X^{\mu}_{cl}\mathscr{D}, P_i) A_{\text{tachyon}}(P_i)(1 + O(\alpha')). \tag{3.3}$$

This is a remarkable result, which hints at a very large and unusual symmetry of string theory that might be restored at high energies. Similar relations could surely be derived for the superstring and could also be extended to multiparticle amplitudes as most of the above story goes through for these. If this is the case then the full S-matrix of the $\alpha' \to \infty$ limit of string theory (say for the heterotic string) could be expressed in terms of the dilaton S-matrix. One could even contemplate constructing explicitly the $\alpha' \to \infty$ limit of the theory by plugging these relations into the unitarity equations (which according to the arguments of Gross & Mende (1988) might be valid for the $\alpha' \to \infty$ limit because the high-energy behaviour is dominated by amplitudes in which all internal momentum transfers and energies are large), and using these to solve for the dilaton amplitudes and thereby the full $\alpha' = \infty$ theory. This approach is made quite complicated by the accumulation of an infinite number of particles at zero mass as $\alpha' \to \infty$.

The above relations connect amplitudes involving particles of different and arbitrarily high spin. If they are generated by a symmetry transformation of the $\alpha' \to \infty$ theory it must be one whose conserved charges have arbitrarily high spins. This would contradict the Coleman–Mandula theorem (Coleman & Mandula 1967), which limits the maximal spin of a conserved charge to be one. Perhaps the Coleman–Mandula theorem is invalid for the $\alpha' \to \infty$ limit of string theory. One of the assumptions of the theorem is the *particle-finiteness* assumption (Coleman & Mandula 1967), which states that for any finite M there are only a finite number of particles with mass less than M. For $\alpha' = \infty$ there are an infinite number of massless particles, in which case the theorem need not apply.

It might also be the case that the Coleman–Mandula theorem is valid. The theorem states that if there exist higher spin conserved charges than the S-matrix equals the identity and the theory is trivial. This is because a higher spin conserved charge, $Q_{\mu_1, \mu_2, \mu_3, \ldots}$, would take values equal to $\sum_i P^i_{\mu_1} P^i_{\mu_2} P^i_{\mu_3} \ldots$ on asymptotic states of spinless particles of momenta $P^1, P^2, P^3 \ldots$. The only way this can be conserved is if all the individual momenta are conserved, i.e. only forward and backward scattering is allowed. Then, if one accepts the usual analyticity of scattering amplitudes, no scattering at all is allowed (in more than two space-time dimensions). Perhaps this theorem is valid, the higher spin symmetries do exist and consequently the scattering amplitudes do all vanish as $\alpha' \to \infty$. This is certainly suggested by the exponential fall-off of the scattering amplitudes, order by order in perturbation theory, as in (2.3), as $\alpha' \to \infty$.

What does it mean to say that the high-energy phase of the theory has vanishing S-matrix? It clearly implies the existence of an enormous symmetry of the theory, since the individual momenta of each string mode are separately conserved. From this point of view the maximal symmetry of the theory is so powerful as to render the theory trivial. *The non-trivial dynamics of the string is then a consequence of symmetry breaking.*

This idea bears some resemblance to Witten's ideas concerning the high-energy phase of gravity, wherein the metric vanishes, physics becomes ultralocal and only topological observables are measurable.

In this case the relations (3.2), although of the form 0/0, would still have much content. They might be expressions of the remaining symmetry in the presence of a non-vanishing M_{Planck}. Or they might be consequences of the simple nature of the symmetry breaking interactions.

In any case these relations hint at a very unfamiliar and new kind of symmetry which, to date, we do not understand. This is perhaps not too surprising as it is very likely that the high-energy phase of string theory is a very unfamiliar phase. Indeed I shall argue that at high energies our standard picture of space and time must break down.

4. CONSTRUCTING A STRING MICROSCOPE

The traditional method of exploring short distances is to scatter particles at ever increasing energies. Our most energetic accelerators function as our most sensitive microscopes. This is because the obstacle to perfect spatial resolution is quantum mechanical fuzziness, as expressed by the Heisenberg uncertainty principle; $\Delta X \approx \hbar/P$. As the momentum of our beams increases we can resolve shorter and shorter distances and probe the very concept of the space-time continuum. With the aid of existing high-energy accelerators we have explored distances down to 10^{-16} cm with no sign of any breakdown in our concepts of space and time. If we were to probe down to distances of order the Planck length, however, it is very likely that some modification of current concepts will be necessary. Many people, in considering the notion of the space-time continuum in the context of quantum gravity where the metric itself fluctuates, have speculated about notions of 'space-time foam', which might alter the character of space-time at the Planck length, or advanced the idea that there is a cut-off (such as a lattice cut-off) below which the continuum loses its meaning.

We now possess a consistent theory of quantum gravity, string theory, in which this question can be addressed in a quantitative fashion. With the aid of the semi-classical methods developed here I shall attempt to explore this issue, to construct a string microscope and explore whether space and time, as defined by operational stringy methods, remains a valid concept to arbitrarily 'short distances'. String theory possesses no cut-off at short distances, Poincaré invariance does not break down and we therefore may contemplate experiments at arbitrarily high energies. None the less, I shall argue that, because of the non-local nature of stringy probes, there is a minimal distance, of order the Planck length, below which we cannot actually resolve distances.

In scattering experiments we do not measure distances directly. We measure only momenta from which we indirectly infer the spatial structure of the scattering event. Thus, tests of the nature of space-time inevitably involve much theoretical analysis. The claim that we have tested the point-like nature of quarks and leptons to distances of order $l \approx 10^{-16}$ cm is based on arguments that if the *theory* is modified at momenta of order Λ then experiment requires that Λ be greater than $\hbar/l$. I do not know of a direct way to tell whether string theory will truly require a modification of the notions of space and time at short distances.

Let us use the semi-classical description of string scattering at high energies to examine whether strings can be used to construct a microscope that would allow us to probe arbitrarily short distances. This is straightforward because we have seen that the scattering in this limit is dominated by a classical trajectory, so we can imagine translating our analysis directly into a space-time picture. Of course the analysis that I reviewed was based on perturbation theory. We therefore must pretend that this perturbation theory is a good approximation. This is tricky because I argued above that the expansion is always divergent. Nevertheless, for the perturbative vacua about which we are expanding, the coupling, g, which is proportional to the gauge or gravitational coupling, is a free parameter. In reality it is given by the expectation value of a dynamical field (the dilaton), but this is not fixed in perturbation theory. Therefore let us take it to be very, very small so that we can trust perturbation theory, at least in the sense of an asymptotic expansion.

The high-energy scattering of the strings was described by a specific surface given by equation (2.10), where z runs over the Z_N surface described previously. It is not too difficult to picture the scattering event described by this formula. For this purpose it is useful to choose a *gauge* where $X^0(\sigma, \tau) = \tau$ and to describe the trajectories $X^i(\sigma)$ as a function of the time τ. For present purposes it is enough to note that the size of the strings at the time of collision, which defines the size of the interaction region or the size of the distances that we can explore using these probes, grows with increasing energy as

$$X_{\text{coll}} \approx \alpha' E/N, \tag{4.1}$$

where E is a characteristic energy of the scattering. This is unfortunate, given that we are trying to use strings as local probes, because we find that as we increase the energy to probe shorter distances, the effective size of the strings increases.

There is one problem with this conclusion, namely the saddlepoint trajectory is imaginary (i.e. note the i in (2.10)). What does this mean? What it appears to mean is that in the high-energy limit of the first-quantized theory the scattering is an exponentially suppressed process that can only be calculated by analytically continuing to imaginary time. This is not a tunnelling process, after all the scattering is certainly allowed as a real process. (For example the velocity flow $dX'(\sigma, \tau)/dX^0(\sigma, \tau)$ is real and physical.) It is analogous to a particle that backscatters, with an exponentially small amplitude, at energies way above a potential barrier.

This remark might be related to our previous conjecture that the symmetry revealed at high energies is the maximal symmetry of no interaction. Indeed, the only way the strings scatter is by a classically unallowed process that we obtain by analytical continuation to imaginary X^μ!

None the less, even if the trajectory described above is not the real Minkowski trajectory, it should describe averages of the real trajectories and thus the average size of the interaction region. At low energies (compared with the Planck mass) we gain in spatial resolution as we increase the energy of our microscope. However, it appears that when we get to the Planck length things change. The fact that our probes themselves are non-local strings becomes relevant. In fact the strings themselves expand with increasing energy. This contributes to the fuzziness with which we can resolve distances, so that the total fuzziness is

$$\Delta X \approx \hbar c / E + G_N E / g^2 c^5. \tag{4.2}$$

Here we have used the fact that (in the heterotic string theory) the string scale (set by α') and the Planck scale are related by g^2. (We can think of g^2 as the fine structure constant and G_N is Newton's constant of gravity.) This is fortunate, as it means that with our assumption of small g^2, the strings never get within their Schwarzchild radius. Thus we do not have to come to grips with strong gravitational effects. The above smearing is not caused by strong gravity, rather it results from the non-local nature of string probes. It implies that the minimal length that can be probed with strings is of order $1/M_{\text{Planck}}$! If this argument is valid then space-time could have no physical meaning below the Planck length.

I expect that in the final formulation of string theory that space and time will emerge only as an approximate concept, valid or useful for certain approximations to the theory. Thus, for example, many people believe that the proper description of the configuration space of string theory is something like the space of all (cut-off) two-dimensional field theories. Conformal field theories are just the classical solutions of the theory. For these we have a space-time interpretation of the perturbative expansion of string scattering amplitudes about these classical vacua. However, for a general two-dimensional field theory there is no such interpretation. (If this is the case then euclidean gravity is probably on shakier grounds than one might otherwise suspect, because it does not make much sense to analytically continue an approximation.) In such a formalism the space-time manifold is not a primary concept, rather it is a useful description of string physics at low energies and weak coupling (where, presumably, perturbation theory is justified).

Further elaboration of these ideas, along with the practical applications of string theory, awaits the development of a non-perturbative formulation of the theory.

This research was supported in part by NSF grant PHY80-19754.

References

Coleman, S. & Mandula, J. 1967 *Phys. Rev.* **159**, 1251.
D'Hoker, E. & Phong, D. H. 1989 Princeton preprint PUPT-1054. *Rev. mod. Phys.* (In the press.)
Gross, D. J. 1988 *Phys. Rev. Lett.* **60**, 1229.
Gross, D. J. 1989 Kniznik Memorial volume. (In the press.)
Gross, D. J. & Mende, P. F. 1987 *Phys. Lett.* B **197**, 120.
Gross, D. J. & Mende, P. F. 1988 *Nucl. Phys.* B **303**, 107.
Gross, D. J. & Periwal, V. 1988 *Phys. Rev. Lett.* **60**, 2105.

Discussion

J. R. Ellis, F.R.S. (*Theory Division, CERN, Geneva, Switzerland*). The contributions to fixed-angle scattering that Professor Gross discussed fall off more slowly with higher energy when one goes to higher genus. Does this mean that the true behaviour of the theory in this limit is controlled by some 'infinite-genus' configurations that we do not understand?

D. J. Gross. Perhaps; it is very difficult to make sense out of a sum of terms that behaves as $g^n n! e^{-A/n}$ in nth order. Were it not for the $n!$ the sum could converge (if g is small enough), but the ubiquitous $n!$ renders the sum divergent. For this reason I have tried to extract from the study of the high-energy behaviour consequences that can be derived order by order, and do not require summing all orders of perturbation theory.